Photovoltaics from Milliwatts to Gigawatts

From the rising of the sun to the place where it sets, the name of the Lord is to be praised.
Psalm 113 v3 (NIV)

Photovoltaics from Milliwatts to Gigawatts

Understanding Market and Technology
Drivers toward Terawatts

Tim Bruton
TMB Consulting
Woking, UK

This edition first published 2021
© 2021 John Wiley & Sons Ltd

The right of Tim Bruton to be identified as the author of this work has been asserted in accordance with law.

Registered Offices
John Wiley & Sons, Inc., 111 River Street, Hoboken, NJ 07030, USA
John Wiley & Sons Ltd, The Atrium, Southern Gate, Chichester, West Sussex, PO19 8SQ, UK

Editorial Office
The Atrium, Southern Gate, Chichester, West Sussex, PO19 8SQ, UK

For details of our global editorial offices, customer services, and more information about Wiley products visit us at www.wiley.com.

Wiley also publishes its books in a variety of electronic formats and by print-on-demand. Some content that appears in standard print versions of this book may not be available in other formats.

Library of Congress Cataloging-in-Publication Data

Names: Bruton, Tim, 1946– author.
Title: Photovoltaics from milliwatts to gigawatts : understanding market
    and technology drivers toward terawatts / Tim Bruton.
Description: Hoboken, NJ : Wiley, 2021. | Includes bibliographical
    references and index.
Identifiers: LCCN 2020030557 (print) | LCCN 2020030558 (ebook) | ISBN
    9781119130048 (cloth) | ISBN 9781119130055 (adobe pdf) | ISBN
    9781119130062 (epub)
Subjects: LCSH: Silicon solar cells. | Photovoltaic power
    generation–History. | Photovoltaic power systems. | Solar energy
    industries.
Classification: LCC TK2963.S55 B78 2021 (print) | LCC TK2963.S55 (ebook) |
    DDC 621.31/244–dc23
LC record available at https://lccn.loc.gov/2020030557
LC ebook record available at https://lccn.loc.gov/2020030558

Cover design: Wiley
Cover image: © alexsl / Getty Images

Set in 10/13pt STIXTwoText by SPi Global, Pondicherry, India
Printed and bound by CPI Group (UK) Ltd, Croydon, CR0 4YY

10 9 8 7 6 5 4 3 2 1

For my grandchildren:
Bethan, Benjamin, Carys, Daniel, Isaac, Jemima, Joseph, Joshua, Samuel

Contents

Preface

Harnessing the power of the sun has been a challenge to the human race for thousands of years. In the last quarter of the twentieth century and into the twenty-first, the deployment of photovoltaic solar energy conversion assumed great importance. In its early years, it promised security of energy supply and price stability for many countries, and today it is a reliable route to carbon-free energy production, a vital tool in the fight to halt manmade climate change. I have been highly privileged in being able to see this dramatic story evolve. I joined the embryonic BP Solar research team in August 1983 when the world market was a mere 20 megawatts and the future uncertain. In the subsequent 37 years, I have seen the technology grow from those early days when many applications were at the milliwatt scale to now, when single photovoltaic power stations of over 1 GWp are in operation. That this has happened is due to the interactions of many actors from the fields of science, manufacturing, and politics.

My ambition in this book is to relate how solar cells transitioned from a high-cost, small-volume, niche market to the global status they enjoy today. The text follows the developments in fundamental understanding, cell processing, and scale-up of manufacturing (and subsequent drop-off in product cost). I tend to use the conversion efficiency of solar cells from solar radiation to useful electricity as a criterion for assessing the progress of particular technologies, but with the caveat that high efficiency must be achieved cost-effectively. Throughout the period in question, I was principally involved at BP Solar and then at NaREC in silicon solar cell design and manufacturing, although along the way I also had responsibilities for thin-film silicon, cadmium telluride, III-V cells, and concentrators. I have taken the opportunity to detail progress in these areas, but also to give my own explanation for why they have not been able to displace silicon as the prime solar cell material. The opinions expressed in this book are entirely my own and are not the views of any of my previous employers.

Throughout my career, I had the privilege of working with many able and dedicated people. In particular, I am indebted to Nigel Mason, Stephen Roberts, Daniel Cunningham, Keith Heasman, and Stephen Ransome for their steadfast support and friendship through the years. Following the merger of BP Solar with Solarex in 2000, I was able to connect with a different generation of pioneers, and it was inspiring to be able to share ideas with Steve Shea, John Wohlgemuth, and the late David Carlson. I have also enjoyed interactions with many others through supporting manufacturing in the United Kingdom, Spain, India, Australia, and the United States. There are many other colleagues from my time in BP Solar and NaREC who helped in innumerable ways, and space does not allow me to acknowledge them all by name. I must thank my wife, Margaret, who selflessly supported me throughout my career and without whom this book would not have been written.

1

The Photovoltaics: The Birth of a Technology and Its First Application

1.1 Introduction

'For more than a generation, solar power was an environmentalist fantasy, an expensive and impractical artefact from the Jimmy Carter era. That was true right up to the moment it wasn't' [1]. This quotation neatly encapsulates the theme of this book: how a technology grew from a high-cost product in a specialist application to a global technology supplying a significant proportion of the world's electricity against a background of at best scepticism and at worst open hostility. In 2018, 102 GWp of photovoltaic modules were installed globally, leading to a total installed capacity of 509 GWp, while an independent study showed that photovoltaics was the lowest-cost means of generation of new-build electricity-generating capacity, including nuclear and fossil fuel sources [2]. At the end of 2019, photovoltaics provided 3% of the global electricity supply, but the expectation is that this percentage will continue to rise until it is the dominant electricity-generating technology by 2050, with 60% of global output [3,4]. Figure 1.1 shows the expected growth of all generating technologies to 2050.

This dramatic development of photovoltaic installations has been the work of many inspired individuals. Their stories are told in other places [5–7]. The aim of this book is to describe how the technology changed from small-area solar cells of 10% efficiency conversion of sunlight to electricity to the mass-production cells of today, with efficiencies in the range 20–24%, and the route to >30% becoming clear. The present chapter describes how the potential for photovoltaic conversion was first recognised and how it moved into the early stages of commercialisation as a high-technology product for use in powering space satellites. Later chapters will describe how this space technology became a terrestrial one and the driving forces and technology developments that made it the global force it is today. Furthermore, the options for going beyond the current technology will be reviewed and the route to achieving terawatt global installations discussed.

It should be no surprise that photovoltaics has achieved the advances it has. Since the invention of the semiconductor transistor in 1948, solid-state electronics has transformed the way in which we live. Computers, mobile phones, the Internet, and so much

Photovoltaics from Milliwatts to Gigawatts: Understanding Market and Technology Drivers toward Terawatts, First Edition. Tim Bruton.
© 2021 John Wiley & Sons Ltd. Published 2021 by John Wiley & Sons Ltd.

Units: PWh/yr

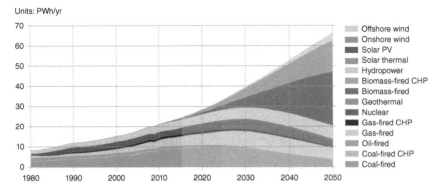

Figure 1.1 Evolution of electricity-generating technologies to 2050 *Source:* DNV GL Energy Transition Outlook 2018

else would not exist without the underlying semiconductor technology. Photovoltaic solar energy conversion is the application of solid-state technology to the energy field. Electricity is generated simply by the absorption of sunlight in a semiconducting diode. There are no moving parts. No liquid or gaseous fuels are needed. There are no effluents requiring disposal and no noise is generated. Sunlight is abundant, delivering to the earth's surface 6000 times humanity's total energy usage [8]. It is the only renewable resource capable of delivering the world's energy needs carbon-free by 2050, and it will remain available for the next 5 billion years. The photovoltaic technology is easily scalable, so that small cells can generate the few milliwatts required for consumer devices such as calculators and watches, while larger ones can be used to assemble modules for deployment at the gigawatt level. It is these advantages which spurred many advocates to continue to promote photovoltaics in the face of significant opposition.

1.2 Sunlight and Electricity

1.2.1 The Early Years

While the potency of the sun has been recognised from ancient times, its role has been mainly that of a source of heat and lighting [9]. It was only relatively recently that the connection between sunlight and electricity was established. Through the nineteenth century, there was an important discovery in this regard about once every decade. Probably the first connection between light and electricity was made by Edmond Becquerel in Paris in 1839 [10]. He observed the flow of an electric current when gold or platinum electrodes were immersed into an electrolyte (acidic or alkaline) and exposed to uneven solar radiation. Some ten years later, Alfred Smee in London observed a current in an electrochemical cell on exposure to intense light, which he called a 'photo-voltaic' circuit – linking the Greek word for light *phos* and the name 'Volta', the original inventor of the galvanic cell [11].

The next step was the observation of photoconductivity in a solid material. A British engineer, Willoughby Smith, in search of a high-resistance metal for use in testing the trans-Atlantic telegraph cable, was recommended selenium. He purchased some selenium rods of between 5 and 10 cm in length and 1 and 1.5 mm in diameter [12]. These were hermetically sealed in glass cylinders, with leads to the outside. They worked well at night, but in bright daylight they became too conducting. Smith concluded that there was no heating effect and that the change in resistance was purely due to the action of light [13]. This stimulated further research into the properties of selenium. The British scientists William Grylls Adams and Richard Evans Day observed current flowing in their selenium sample when no external voltage was applied and were able to show that 'a current could be started in the selenium by the action of light alone' [14]. They had demonstrated for the first time that light caused the flow of electricity in a solid material. They used the term 'photoelectric' to describe their device, and Adams believed it could be used as a means of measuring light intensity [15].

The narrative now switches to America, where Charles Fritts made the first working solar module by covering a copper plate with a layer of selenium and applying a semitransparent gold layer as the top electrode [16]. An example is shown in Figure 1.2. Fritts described the module as producing a 'current that is constant and of considerable force . . . not only by exposure to sunlight but also to dim diffused light and even to

Figure 1.2 Charles Fritts' first photovoltaic array, produced in New York City in 1884 [16] (Courtesy New World Library)

lamplight.' He supplied samples to the German electricity pioneer Werner von Siemens, who greeted them enthusiastically, announcing Fritts' module to be 'scientifically of the most far-reaching importance'. However, its low efficiency – below 1% – made it of little commercial importance. Indeed, there was considerable scepticism at the time, with solar cells viewed as some kind of perpetual-motion machine. The principles of their operation were not understood. One of the leading physicists of the day, James Clerk Maxwell, while welcoming photoelectrcity as 'a very valuable contribution to science', wondered 'is the radiation the immediate cause or does it act by producing some change in the chemical state' [16].

The underlying science of photovoltaics was given a big boost by the parallel discoveries and developments in photoemission. Hertz observed in 1887 that ultraviolet light caused a significant increase in the sparks in an air gap between electrodes and that it was a function of the wavelength of the light rather than its intensity [17]. While a number of physicists worked on the effect, it was Albert Einstein in 1905 who explained it in terms of different wavelengths behaving as particles of energy, which he called 'quanta' but which were later renamed 'photons'. These quanta had different energies depending on their wavelength. Einstein was awarded the Nobel Prize in 1921 for this work [15]. While these discoveries and other advances in quantum mechanics at the start for the twentieth century did not directly explain photovoltaic effects, they did provide a scientific basis for understanding the interaction of light and materials.

Although research continued on developing solar cells, little progress was made. However, photovoltaics still had its advocates in the 1930s. Ludwig Lange, a German physicist, predicted in 1931 that 'in the distant future huge plants will employ thousands of these plates to transform sunlight into electric power . . . that can compete with hydroelectricity and steam driven generators in running factories and lighting homes' [15]. A more pragmatic view was taken by E.D. Wilson at Westinghouse Electric, who stated that the efficiency of the photovoltaic cell would need to be increased by a factor of 50 in order for them to be of practical use, and this was unlikely to happen [15]. Actually, as will be shown in later chapters, a factor of 20 was achieved, and this was sufficient to create the current global markets.

While progress in other areas of technology was immense in the nineteenth and early twentieth centuries, little real advancement in photovoltaics had been made since Becquerel's discovery a hundred years previously. Entering into the second half of the twentieth century, everything would change.

1.2.2 The Breakthrough to Commercial Photovoltaic Cells

It is well known that the birth of the commercially successful photovoltaic cell dates back to April 1954, when Pearson, Chapin, and Fuller demonstrated the first 6% efficient cell using a p/n junction in silicon. It is no surprise that this discovery occurred at Bell Telephone Laboratories, which was one of the world's premier research laboratories until its forced break-up in 1984. As the research arm of the American Telephone and Telegraph Company, it had a long history of successful innovation, with nine Nobel Prizes awarded over time for work done there. Perhaps its most notable success was the demonstration in 1948 of the point-contact germanium transistor. This illustrates the strength and depth of both the theoretical understanding and expertise in semiconductor processing at Bell labs [18].

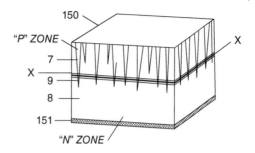

Figure 1.3 Ohl's patented solar cell structure [20] *Source:* R.S. Ohl: US Patent Application filed 27th May 1941

Russel Ohl, a Bell Labs scientist interested in exploring the crystallisation of silicon, is recognised as the discoverer of the p/n junction in this material, in 1941 [19]. In directionally solidifying 99.85% pure silicon, Ohl noted a change in the structure of the solidified ingot, with the upper portion becoming columnar and the lower portion showing no structure; a striated region appeared between the two, forming a barrier to conduction. The upper zone was p type while the lower zone was n type [20]. This can be easily understood as a result of the segregation of dopants during the crystallisation process. While measuring the resistance of rods containing the barrier, Ohl noted a sensitivity to light, which he termed a 'photo electromotive force'. He proceeded to patent this as a solar cell, although its efficiency was similar to that of the selenium cells, at about 1% [20]. Figure 1.3 shows Ohl's silicon structure, the n type region being fine-grained crystallites and the cell contacts plated rhodium. The low efficiency is not surprising given the relatively impure starting material, its multicrystalline nature, and the fact that the n type region was 0.5 mm thick. The relatively low efficiency meant little further work was done until a new approach at Bell Labs.

Success came in the 1950s. The first transistor had been demonstrated at Bell in 1948 using germanium, and had entered commercial production in 1951 [21]. However, germanium had some disadvantages in its fragility and stability, and silicon offered a better option – although a working silicon transistor was not demonstrated until 1954. Two scientists working on this were Calvin S. Fuller and Gerald L. Pearson. Fuller was an expert in doping silicon, while Pearson was an experimentalist. There were three iterations before a good working solar cell was demonstrated [22]. Initially, while not looking for a solar cell, Fuller produced a p type gallium-doped silicon sample, which Pearson dipped into a lithium bath to form a shallow n type region. When Pearson exposed the sample to light, he found to his surprise that a current was generated. At the same time, in a different department, another scientist, Daryl M. Chapin, was looking for a power source for telecommunications repeaters in hot humid locations where conventional dry cell batteries rapidly failed. Chapin concluded that solar cells were a good option, but his experiments with commercial selenium cells of low efficiency were disappointing. He and Pearson knew each other, and Pearson offered Chapin his lithium-doped 'solar cell'. Chapin tested it and found it 2.3% efficient – an enormous

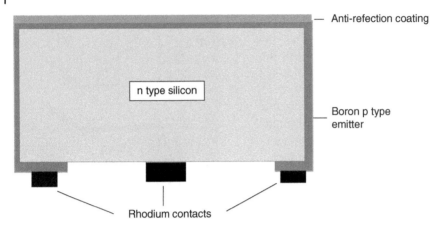

Figure 1.4 Schematic of the first successful silicon solar cell

improvement on selenium, justifying further investigations into silicon's potential. The next step was to replace the lithium with phosphorus. A small amount of phosphorus was evaporated on to the p type silicon to make a shallow n type region. Initial results weren't particularly good, but then Chapin applied a thin plastic layer to act as an antireflection coating (ARC) on the otherwise highly reflecting silicon surface. This gave the encouraging result of around 4% cell efficiency, which was good progress toward Chapin's target of 5.7% for a viable power source. However, further progress was slow, and forming a good electrical contact proved to be an ongoing problem. A spur to further activity came from Bell's competitor, RCA, which was developing an 'atomic battery' using a strontium 90 source to irradiate a silicon solar cell, although efficiencies were poor. The breakthrough came when Fuller, who had been experimenting with boron to give a p type silicon emitter which offered a new configuration, demonstrated that heating an n type silicon wafer for 5.5 hours at 1000 °C in a boron trichloride atmosphere under reduced pressure could produce a 0.25 µm-deep diffusion with a resistivity of .001 Ω/cm [23]. This equated to 40Ω per square sheet resistance emitter, which is a typical figure for later commercial silicon solar cells. Arsenic was used to dope the silicon base n type to 0.1 Ω/cm, and this was then cut into long narrow strips in accordance with the best previous cell results. The emitter was formed using the new boron diffusion process, with the emitter itself wrapping around the cell as shown in Figure 1.4. It was then partially removed on the rear to expose the n type base. Contacts were made by electroplating rhodium to the exposed base and emitter [24]; the relative ease in forming these contacts represented a significant advance. A polystyrene layer with refractive index 1.6 was used as the ARC. The solar cell efficiency was approximately 6% [25]. Chapin had proposed a theoretical limit of 22% efficiency for silicon, but had stated that practical limitations would result in its being lower in the real world. The spectral response of the cell is shown in Figure 1.5. It was proposed that 10 cells

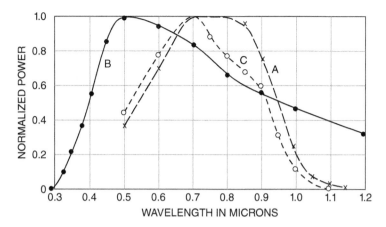

Figure 1.5 (A) Normalised spectral response of a p on n silicon solar cell. (B) Solar spectrum. (C) Relative integrated response [23] *Source:* C.S. Fuller: US Patent 3015590, 2nd Jan. 1962 (filed 5th March 1954)

connected in series would be needed to provide the 1 W of power required to charge the batteries at a repeater station in a rural carrier telephone system [24].

The 6% efficiency result met Chapin's original target and gave the impetus for further improvement. By November 1954, an 8%-efficiency cell had been produced with the same wraparound structure [26], and by May 1955, 11% efficiency was demonstrated [27]. It is interesting to note that this structure – an n type wafer with all of the contact metallisation on the rear – has a resonance with the current world-record silicon solar cell (26.7% efficiency), based on an n type wafer with rear contacts, but with amorphous silicon providing the p side of the junction [28]. In their 1954 patent filing, the inventors stated that 'Sunlight is the most common, most accessible, and most economical form of energy on the Earth's surface' [24]. This group at Bell Labs opened the door for practical exploitation of this rich energy source, although it was a further 20 years before its true potential began to be fully exploited.

1.2.3 Hiatus

While the Bell discovery was heralded as a great breakthrough, with the *New York Times* stating it 'may mark the beginning of a new era, leading eventually to the realisation of one of mankind's most cherished dreams – the harnessing of the almost limitless energy of the sun for the uses of civilisation' [29] and the *US News & World Report* claiming 'the (silicon) strips may provide more power than all the world's coal, oil and uranium' [30], the reality in 1954 was somewhat different. The cost of the high-purity silicon needed to make viable solar cells was very high at $845/kg, resulting in an estimated cost of electricity of $144/kWh, compared to $23.7/kWh for dry cell batteries and less than $0.02/kWh for retail grid electricity [31]. Despite this disadvantage, Bell's manufacturing subsidiary, Western Electric, took up commercialisation.

One of the first applications was powering a remote telephone line in rural Georgia (USA). However, the introduction of silicon transistors for voice amplification with a very low power requirement made the application redundant and activity at Bell largely declined [31]. Western Electric licensed other companies to manufacture the silicon solar cells, one of which, National Fabricated Products, went as far as to assemble a range of demonstration products. But no market was forthcoming and National Fabricated Products was taken over by Hoffman Electronics in 1956. Hoffman provided prototypes for both the US Coast Guard and the US Forest Service, but no business resulted. The initial enthusiasm cooled to the extent that in 1957 it was reported that 'Viewed in the light of the world's power needs, these gadgets are toys' [32].

1.2.4   The First Successful Market: Satellites

Although a terrestrial market was slow to materialise, Hoffman Electronics continued to do research to improve solar cell efficiency and reduce cost. By the end of 1957, it had demonstrated a 12.5% ($1\,cm^2$)-efficient silicon cell [33]. The discoveries at Bell Laboratories did not pass unnoticed amongst the military, and the US Army Signal Corps visited the company to evaluate the technology and concluded that the only viable application was for the power supply of an artificial earth satellite, which was a top secret project at the time [34]. This view was shared by the US Air Force – but not by the US Navy, which was eventually awarded the satellite project, and which had decided that silicon solar cells were 'unconventional and not fully established' [34]. Interdepartmental rivalries also played a part [35]. Nonetheless, when President Eisenhower first announced the satellite programme publicly in 1955, the *New York Times* published a sketch showing solar cells providing the power. Intense lobbying was carried out, particularly by Dr Hans Ziegler of the Signal Corps Research and Development Laboratory (USARDL), assisted by a decision to simplify the satellite's mission and lighten its payload. Eventually, it was agreed that both a dry cell battery power supply and a solar-powered transmitter would be used. Solar cells were tested by attaching them to the nose cones of two high-altitude rockets and were shown to survive the rigours of the launch and of the space environment.

The first launch in December 1957 of the Vanguard TV-3 satellite, depicted in Figure 1.6, failed, but the satellite itself was recovered [36]. The Vanguard 1, identical to the TV-3, was successfully launched on 17 March 1958, with the expectation that it would remain in orbit for 2000 years (although later recalculation lowered this to 240 years). It is the oldest artificial satellite still orbiting the earth. Its scientific purpose as part of the International Geophysical Year was to obtain geodetic measurements of the shape of the planet. As predicted, the battery-powered transmitter stopped working in June 1958, while the solar-powered one continued until May 1964, when the last signals were received in Quito, Ecuador. The satellite was a 16.5 cm aluminium sphere weighing 1.48 kg. A 10 mW 108 MHz telemetry transmitter was powered by a set of mercury batteries, while a 5 mW 108.03 MHz beacon transmitter was powered by solar

Figure 1.6 Vanguard TV-3 satellite on display at the Smithsonian Air and Space Museum. Recovered from its failed launch on 6 December 1957 [36] *Source:* Smithsonian National Air and space Museum (TMS A19761857000cp02)

cells mounted in six 5×5 cm arrays, which gave a nominal power of 1 W, equating to an efficiency of 10% at 28 °C [34]. The cells were supplied by Hoffman Electronics.

This proved to be a watershed moment for photovoltaics, as the technology had proved itself a reliable long-term source of electricity in a real application.

1.3 Photovoltaics Demonstrates Success

With the highly visible success of the Vanguard 1 mission, photovoltaics became the technology of choice for powering satellites. Nevertheless, some scepticism remained in place, and photovoltaics was seen purely as a stopgap measure until atomic batteries were developed. Others thought that while photovoltaics was acceptable for the simple early satellites, it would not provide enough power for the more sophisticated ones – not to mention space stations – envisaged for the future [37].

Immediate confirmation of the value of solar cells in space was given by the Russians, who launched the Sputnik 3 on 5 May 1958, also as part of the International Geophysical Year. In addition to a number of scientific instruments, Sputnik 3 included a solar cell-powered transmitter [38]. The Russians had been monitoring Bell Laboratories' work and had developed their own cells [37]. One Russian scientist stated in 1958 that 'Solar batteries . . . would ultimately become the main source of power in space'. With hindsight, it is easy to see how these predictions became true. Space was the obvious market for solar cells. Solar insolation is 40% higher in space than on the earth's surface. Satellites are the ultimate off-grid market, with no maintenance possible after launch until at least the 1990s. They are free of the day–night constraints of terrestrial systems. Price is not an issue, given the high cost of launchings satellites. In any case, with increasing production volume and the general growth of the silicon semiconductor infrastructure, solar cell costs had fallen to $100/W by 1970.

Space offered some challenges, however. It had been noted by RCA Laboratories that in its experiments to produce electro-voltaic cells, beta radiation from a Sr90-Y90 source had seriously degraded solar cell performance [39]. Initially, figures for cosmic radiation were extrapolated from high-altitude balloon flights, and it was calculated that it would take 105 years for there to be a 25% loss of cell efficiency. However, the Explorer 1 satellite, launched in 1958, found that in the yet to be recognised van Allen radiation belt, cosmic radiation was 104 times higher than that estimated from balloons, meaning unshielded solar cell life would be only 10 years [39]. This was mitigated by the discovery that a quartz or sapphire cover could significantly reduce the amount of radiation reaching a cell, prolonging its active life. Nevertheless, radiation resistance of solar cells became a major topic of research in subsequent years, and end-of-life cell efficiency (rather than efficiency at launch) a major consideration. Research at the USARDL found that reversing the structure of the original Bell Labs cells by producing a shallow n doped layer via phosphorus diffusion into a p type substrate increased radiation resistance by an order of magnitude [40]. As a result, n on p solar cells became the preferred structure for use in space. It was also much easier to create shallow diffusions with phosphorus than with boron.

The mid 1950s proved a fertile time for understanding of the p–n junction solar cell. Van Roosbroeck and Pfann predicted in 1954 that such a cell could reach 18% efficiency [41]. A later paper by Prince calculated junction depths and series resistance and concluded that 21.7% efficiency was the maximum possible in silicon, but that 10% was a practical limit [42]. This analysis enabled Bell Labs to move forward from its 6% cell to a 10% one, although incorrect data had been used for the absorption depth of solar photons. Prince also concluded that the optimum bandgap would lie between 1.0 and 1.6 eV. As well as providing the basis for further work in silicon, this also stimulated research into other semiconductors such as cadmium telluride, indium phosphide, and gallium arsenide [39]. Interestingly, a paper by Jackson demonstrated how solar cells of different bandgaps could be stacked on each other and that a tandem cell with bandgaps of 0.95, 1.34 and 1.91 eV would have an efficiency of 37% assuming the cell would operate as well as the single junction cell of that time [43]. These papers paved the way to the definitive work on ultimate solar cell efficiencies for semiconductors of different bandgaps, published by Schockley and Quiesser in 1961 [44]. This ultimate efficiency was modelled by assuming 100% absorption of incoming light and no losses within the solar cell. The main losses were imposed by the limitation of the semiconductor bandgap. Photons with energy less than the bandgap could not excite electrons into the conduction band, and the energy was absorbed as heat. Photons with energy above the bandgap created an electron–hole pair, but with excess energy in the conduction band, which had many empty lower-energy states close to the band edge. The excited electrons lost their energy again by heat and occupied those empty energy states. Schockley and Quiesser found the optimum bandgap was 1.4 eV, giving an efficiency of around 30%. Finding ways to exceed this Shockley–Quiesser limit has been the topic of much research and is discussed in Chapters 7 and 8.

1.3.1 First Commercial Operation

The combination of good results from actual satellite flights and the potential to achieve efficiencies as high as 30% in the long term provided a good platform for the continued use of solar cells in space. In 1959, Hoffman Semiconductor supplied 9600 p on n solar cells, 1×2 cm, to the Explorer 6 satellite. The cells were mounted on four paddles, which deployed from the satellite's equator. Explorer 6 took the first pictures of earth from space [45]. A further step forward was the launch of the first commercial telecommunications satellite, 'Telstar', in 1961. This was a project of Bell Telephone Laboratories, first proposed in 1955 [39]. When launched, Telstar utilised the latest development in solar cells for space [46]. Its cells were made by the Bell subsidiary Western Electric. Radiation resistance was the first parameter to be studied, and the researchers confirmed the earlier observation that n on p cells were 10 times more resistant that the original p on n cells. Their results are shown in Figure 1.7.

The figure shows the clear superiority of the blue-sensitive (i.e. shallow junction) phosphorus-diffused n on p solar cell. The short-circuit current provides a good indicator of the underlying electronic quality of the silicon material in the cell. Further radiation protection was provided by a 750 μm-thick sapphire cover glass. The next objective was to optimise the efficiency of the solar cell. It was found that the highest efficiencies were achieved with silicon wafer doping at 1 Ω/cm. Next, diffusion was optimised: while lighter diffusions improved blue response (with 90 Ω per square being an optimum), the cells were prone to electrical shunting. A final diffusion range of 30–60 Ω per square was thus adopted. Finally, improvements were made to the metallisation. While the original Bell cells had all the metallisation on the rear, Wolf at Hoffman had shown that a gridded pattern on the front increased solar cell efficiency [47]. Five grid lines (150 μm wide) per cell were applied. A schematic of the Telstar cell is shown in Figure 1.8.

As well as adopting a grid structure, the metallisation was changed to give a better contact. The electroplated nickel of the original cell was replaced by an evaporated titanium contact as the nickel plating still gave problems even with the diffusion compromise. On sintering, the titanium reduced the native silicon dioxide coating to make contact with the n type emitter. The titanium contact was coated with silver to prevent oxidation and provide a good surface for soldering. Another parameter to be considered was thermal shock and the range of operating temperatures. Under normal operation, the solar cell temperature would varying between 10 and −50 °C, although at certain times this could extend to 100 to −69 °C . This could cause cracking in the solar cell. It was found necessary on the rear contact to mask the cell edges in order to leave an

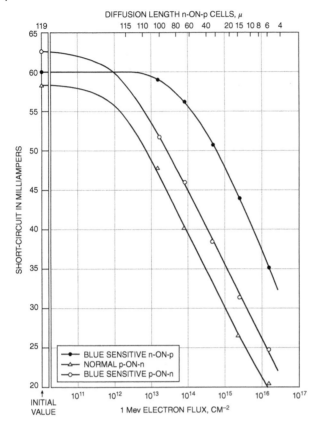

Figure 1.7 Decrease in short-circuit current of n on p and p on n silicon solar cells for different fluxes of 1 MeV electrons [46] (Source *Bell System Technical Journal*)

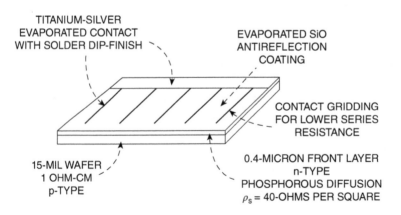

Figure 1.8 Schematic of the Telstar silicon solar cell [46] (*Source Bell System Technical Journal*)

Figure 1.9 Current–voltage curve for a Telstar cell at 28 °C at 100 mW/cm² [46] (*Source Bell System Technical Journal*)

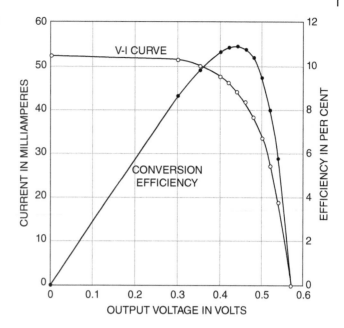

uncoated border to avoid cracking. Two cells 1×2 cm were cut from a 1 inch-diameter silicon wafer. A silicon dioxide antireflection coating was applied by evaporation. In a run of 10 000 cells, the efficiency at 100 mW/cm² insolation, 28 °C, and a fixed load point at 0.45 V had a median efficiency of 10.8%. A typical I/V is shown in Figure 1.9. 3600 cells were used for the satellite, in 50 parallel strings of 72 cells, to give a total power of 14 Wp.

The satellite mission was successful, although radiation damage was extensive and the solar cell output decreased to 68% of its original value after 2 years in orbit. It was calculated that even with the state of the art, additional solar cells could be deployed to give a 20-year life without significant weight penalties.

The Telstar cell provided a basis for further silicon development. The n on p structure was readily manufactured and the diffusion range of 30–60 Ω/cm on a 1 Ω/cm silicon wafer became the standard for terrestrial solar cells for many years. The only significant difference between the Telstar cell and later terrestrial cells was that Telstar used (111) orientation wafers while the later cells used (100) orientation, which could be chemically textured as discussed in the next section.

1.3.2 Continuing Research for Space

The emergence of solar cells as the only viable power source in space, coupled with the competition between the United States and Soviet Union over leadership in the space race, stimulated further research. Between 1958 and 1969, the US government provided $50 million for solar cell research. In 1961, the USARDL demonstrated a 14.5%-efficient solar cell similar to the Telstar cell [48]. But, remarkably, while the industrial production of space solar cells ramped up, their efficiency showed no increase between 1961 and 1970, as illustrated in Figure 1.10 [49].

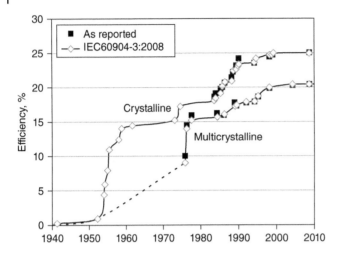

Figure 1.10 Evolution of recorded mono- and multicrystalline silicon solar cell efficiencies, 1940–2010 [49] (Courtesy Wiley) *Source:* M.A. Green: Progress in Photovoltaics-Research and Applications 17 (2009) 183–189

Increasing radiation resistance had been prioritised over increasing solar cell efficiency. Reducing cell cost was also not a major concern. By 1970, attendance at the 8th IEEE PV Specialists Conference had fallen to only 100 delegates. The next milestone after the USASRDL 14.5% cell was the 15.2%-efficient cell reported by Comsat Laboratories in the United States in 1973. COMSAT (Communications Satellite Corporation) was founded in 1962 to develop commercial communication satellites, and in 1969 it established its own research laboratories. In 1972, it reported the first violet cell, which had a very lightly doped emitter 0.1 μm deep with a sheet resistance of 500 Ω/cm, requiring many fine grid lines (30 per centimetre) to minimise series resistance, as well as tantalum pentoxide ARC to allow more blue light into the cell [50]. The initial efficiency reported was 13%, but this was quickly improved to 14% [51]. The cell was shown to maintain its higher efficiency compared with standard commercial cells after 1 MeV electron irradiation. However, it was soon overtaken at COMSAT by the chemically textured cell. This involved silicon wafers textured using dilute alkali etching of (100) orientation silicon. The (100) direction etched faster than the (111), so that pyramids typically 2–3 μm across the base formed on the surface, as shown in Figure 1.11. This greatly reduced reflection, as incident light reflected from one pyramidal surface impinged an adjacent one and so was absorbed. The pyramidal structure also enhanced current collection by refracting the incoming light so that it was more parallel to the collecting p/n junction. COMSAT reported a 15.5% efficiency [51], but this was rapidly increased to 17.2% [52] without loss of radiation hardness.

Another approach to increasing solar cell efficiency involved using higher-resistivity wafers, as these had longer minority carrier diffusion lengths, giving higher short-circuit currents. However, this came at the cost of a lower Voc, so that there was an overall loss in efficiency [53]. It was also desirable to decrease the wafer thickness in order to improve the electrical yield per unit weight – an important consideration for satellites. It was shown experimentally that Voc above 580 mV could be obtained by a

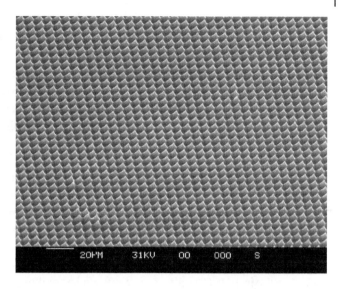

Figure 1.11 Textured (100) monocrystalline silicon wafer. *Source:* Courtesy BP Archive

'gettering process' [54]. Further experiments showed that firing an aluminium rear contact produced good results and that a thickness reduction to 200 μm was possible without loss of efficiency and with good radiation hardness [54]. The aluminium rear surface could also act as back surface reflector of light, contributing to improved current generation. The cells produced were termed back surface field (BSF) cells, and at the standard thickness of 300 μm they increased the solar cell efficiency from 9.3 to 11.6%. The mechanism for efficiency improvement was a combination of the p+ doping at the rear creating an internal electric field that repulsed the minority carrier electrons and a gettering action removing impurities from the silicon introduced in processing to give a high minority carrier lifetime. In later years, the BSF approach represented a useful enhancement in terrestrial solar cells.

This was all demonstrated in the mid-1970s, by which time research for terrestrial applications had begun and further advances in efficiency were being made for terrestrial silicon cells, as described in Chapter 4. Meanwhile, developments in space cells turned toward III–V materials, as described in section 1.4.

1.3.3 Solar Cell Manufacturing

By 1972, there were a thousand satellites in space, with power requirements which varied from milliwatts for Vanguard 1 to kilowatts for Skylab A [55]. All the attendant solar cells had to be manufactured somewhere. Initially, Hoffman Semiconductor and Western Electric supplied some, but once the application had been established, new entrants rapidly appeared, not only in the United States but also in Germany and Japan.

One of the first start-ups in the United States was Spectrolab in 1956, which initially manufactured optical filters and mirrors but in 1958 started in the space cell business, supplying a solar array to the Explorer 1 satellite and then another to Explorer 6 in 1959.

Since then, it has supplied many satellites, including for Syncom, the first successful mission in a geosynchronous orbit. In 1993, the company installed equipment for the manufacture of III–V solar cells, signalling the move way from silicon in space. It currently offers cells of 30% efficiency. In 1975, Spectrolab was acquired by Hughes Aircraft, and in 2000 it became part of Boeing Corporation. It has produced over 4 MWp of space solar cells to date, and it continues in production [56]. At the same time, the Applied Solar Energy Corporation started manufacturing space cells and pioneered the deposition of GaAs cells on germanium substrates. ASEC became Tecstar Inc., but it entered chapter 11 bankruptcy in 2002 and its assets were acquired by Emcore Inc. COMSAT, as already described, started in 1962. In Germany, AEG-Telefunken in Heilbronn started production of space solar cells, with the first supplied to the German AZUR1 satellite in 1968 [57]. The company still produces cells today, after many changes of ownership, under the name Azur Space Solar Power GmbH. It has provided cells for over 400 space projects. In Japan, the Sharp company started research into solar cells in 1959, and specifically addressed the space market from 1967 with the first deployment in space of the Ume satellite in 1976 [58]. Sharp continues today, making triple-junction III–V cells for both space and concentrator solar cell applications.

While manufacture of space cells continues today, it is now a relatively small market compared to the terrestrial one. However, the expertise gained in manufacturing for space provided the basis for the growth of the terrestrial market following the 1973 oil crisis, as discussed in Chapter 2. By 1970, it had already been noted that growth prospects in the space cell industry were limited, and scientists were beginning to look to earth [39]. Sharp Corporation had consistently used reject space cells for terrestrial applications, supplying photovoltaic power to 256 lighthouses between 1961 and 1972 [59]. It demonstrated the first viable silicon module for terrestrial use in 1963 and became a major supplier of photovoltaic modules from 1980 onward. AEG-Telefunken started terrestrial solar cell manufacture at its Wedel site in the mid-1970s [57]. Other 'pure' terrestrial solar cell companies were founded by individuals leaving space cell companies. Bill Yerkes had been president at Spectrolab, but following the Hughes Aircraft acquisition he left the company to form Solar Technology International, which subsequently became ARCO Solar and was the largest photovoltaics manufacturer for several years [60]. Joseph Lindmayer and Peter Varadi left COMSAT in 1973 to form the Solarex company [61], and this too – after a merger with BP Solar – became the world's largest photovoltaics manufacturer, in 2000 [61].

In this way, the development of solar cells for space formed the foundation of the major manufacturing industry and global energy supply that photovoltaics has become.

1.4 Gallium Arsenide and III–V Alloys for Space

Despite its many advantages, silicon also has negative characteristics. Namely, it is an indirect bandgap semiconductor and the bandgap at 1.1 eV is not the optimum. This means that a relatively thick absorber layer is required to absorb the solar spectrum, and the bandgap limits the ultimate efficiency that can be achieved. The semiconductor gallium arsenide (GaAs) does not have these disadvantages. As a direct bandgap semiconductor, it requires only a few microns of absorber, which lowers the mass of the solar cell – an important characteristic in space applications. The bandgap at 1.4 eV is very near the optimum for a single-junction solar cell [44]. GaAs also has the advantage that a wide range of alloys of various group III and V elements exist with different bandgaps and are lattice-matched to it [62], enabling the high-efficiency tandem structures envisaged by Jackson [43]. On the other hand, GaAs is a high-cost technology, with the cost of substrates and active film deposition being much higher than that for silicon. Nonetheless, the cost of the solar cell is a much less important parameter than kWp/kg for space applications [63]. GaAs also has the advantage that it is more radiation-hard than silicon [64]. GaAs cells have been used in space since the 1960s, with the Russian Venera 2 and 3 missions to Venus. In 1986, 70 m^2 of GaAs solar cells were installed on the MIR space station; these functioned for 15 years [65]

The development of III–V cells can be divided into two programmes: single-junction GaAs cells on either a GaAs substrate or germanium and tandem cells usually on a germanium substrate for use in space but also for concentrating solar cell applications. The bandgaps and lattice constants of the most important III–V compounds are given in Figure 1.12. Silicon solar cells were the preferred space cell technology until the 1990s, when III–V cells began to be used. They remain the dominant technology today.

1.4.1 Single-Junction GaAs Solar Cells

GaAs has a history almost as long as that of silicon. One of the first GaAs solar cells was made at RCA Laboratories in 1956, with an efficiency of 6% on very small-area solar cells [66]. From that point on, significant progress was made, driven by the expectation of higher efficiency and enhanced radiation hardness. By 1981, the 20%-efficiency

Figure 1.12 Bandgap and lattice constant for the important III–V alloys [62] (Courtesy Royal Society of Chemistry) *Source:* H. Cotal et al: Energy and Environmental Science 2 (2009) 174-192

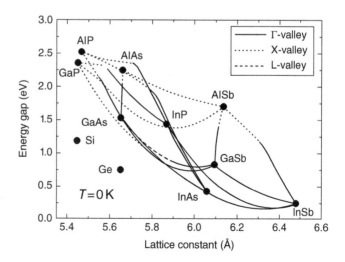

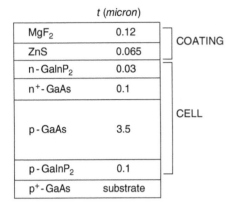

	t (micron)	
MgF$_2$	0.12	COATING
ZnS	0.065	
n - GaInP$_2$	0.03	
n$^+$- GaAs	0.1	
p - GaAs	3.5	CELL
p - GaInP$_2$	0.1	
p$^+$- GaAs	substrate	

Figure 1.13 Structure of a 25.7% GaAs solar cell under the AM1.5 Global spectrum [68] (Courtesy IEEE) *Source:* S.R. Kurtz, J.M. Olsen and A. Kibbler: Proc 21st IEEE PVSC (1990) 138-140

barrier had been breached, with an n+/p/p+ GaAs structure on both germanium and GaAs substrates [67]. Early GaAs cells were made by liquid-phase epitaxy, which had limits in terms of the alloys that could be produced, while the 1981 work was performed using vapour-phase epitaxy. By the late 1990s, however, metal organic chemical vapour deposition (MOCVD) had become a well-established technique for producing a wide range of III–V compounds [64]. In 1990, cells of over 25% efficiency were demonstrated using MOCVD for the epitaxial layers – in some cases, as thin as 0.1 μm [68]. The best cell had an efficiency of 25.7% under the AM1.5 Global spectrum; its structure is illustrated in Figure 1.13. The improved efficiency was derived in part from the use of a GaInP$_2$ window at front and rear, such that the active carrier collection region was separated from the high recombination surfaces at the front and rear of the active GaAs cell.

GaInP$_2$ was used in preference to the GaAlAs$_2$ previously employed, as this was prone to degradation by the inclusion of oxygen. The work highlighted that the electronic quality of the individual layers was as important as the overall device structure in achieving very high efficiencies. Development of single-junction GaAs has been relatively slow, as more research has gone into the higher-efficiency potential triple-junction cells. In 2008, the record efficiency was 26.1% [69]; in 2018, Alta Devices reported a new record for single-junction GaAs cells of 28.9% [70], in the form of an ultrathin cell. This was intended not for space applications but for terrestrial ones, where very high efficiency is important (e.g. the Internet of Things, unmanned aircraft).

1.4.2 Multijunction Solar Cells for Space

The first successful GaAs heterojunction solar cell was demonstrated in 1970 for a GaAlAs/GaAs structure [71]. Good progress was made, and by 1981 an AM0 efficiency of 15% was posted for a GaAs/GaAlAs tandem [72]. At the same time, it was observed that germanium could also be used as a substrate, as it is also lattice matched to GaAs, as shown in Figure 1.12 [67]. At this point in time, liquid-phase epitaxy was the preferred method for making the tandem cells. Lattice matching is important as if a significant mismatch occurs, the stress at the interface between two semiconductors creates threading dislocations which spread through the epitaxial layers and act as recombination centres. These have a great impact in reducing the

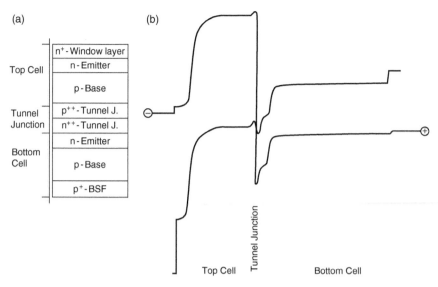

Figure 1.14 (a) Structure of a tandem cell, showing the tunnel junction. *Source:* N.J. Ekins-Daukes: in "Solar Cell Materials-Developing Technologies" ed G J Conibeer and A Willoughby. Pub J Wiley (2014) 113-143 (b) Associated band diagram, highlighting the tunnelling effect [73] (Courtesy Wiley) *Source:* N.J. Ekins-Daukes: in "Solar Cell Materials-Developing Technologies" ed G J Conibeer and A Willoughby. Pub J Wiley (2014) 113-143

solar cell efficiency. Removing the lattice-matching constraint means a wider range of III–V compounds can be used, with greater potential for high efficiency. In this case, alternating layers with different lattice parameters of stress and compression can prevent the generation of threading dislocations. These are known as metamorphic cells [64]. In general, space cells have been lattice matched, but metamorphic cells have been developed for concentrator applications. A particular challenge with the use of tandem cells is that there is a p/n junction between the top cell and the bottom one, which is rectifying for the direction of current flow. To overcome this, a heavily doped 'tunnel junction' is used, in which carriers can travel through the rectifying junction, as illustrated in Figure 1.14 [73]. The requirements of the tunnel junction are that it be a wide-bandgap semiconductor for good transparency and that it be heavily doped in order for tunnelling to occur. This makes for an added complexity and cost for tandem cells.

As already described, the availability of MOCVD deposition systems greatly assisted the development of III–V epitaxial growth. By 1995, the practical limit for a GaInP$_2$/GaAs tandem cell of 30% was reached [74]. At the same time, efforts were underway to fabricate III–V cells for use in space. Germanium was preferred to GaAs as the substrate, as it was physically much more robust. Initially, it was only used as an inactive substrate, but its low bandgap at 0.7 eV made it ideal for use as the bottom cell in triple-junction tandem. Initial production of single-junction GaAs/Ge cells was reported in 1990, with AM0 spectrum cells achieving 18% efficiency for a 4×2 cm cell [75].

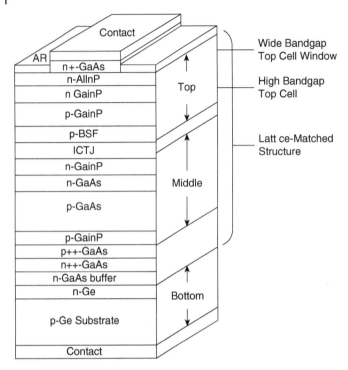

Figure 1.15 Structure of a very high-efficiency triple-junction tandem III–V solar cell [78] (Courtesy Wiley) *Source:* H. Yoon et al: Progress in Photovoltaics-Research and Applications. 13 (2005) 133–139

By 1996, double cells were being made with GaAs/Ge, and a best cell with an efficiency of 23.8% was reported [76]. At the same time, results were being published for triple-junction GaInP$_2$/GaAs/Ge cells with an efficiency of 25.7% (4 cm^2) [77]. The first commercial satellite with dual-junction cells was launched in 1998, with initial cell efficiencies of 22%. By 2005, triple-junction cells with an initial efficiency of 28.0% (AM0 spectrum) were being manufactured [78]. The structure of a triple-junction solar cell is depicted in Figure 1.15, which shows the complexity of the device structure. The current best efficiencies being manufactured for use in space belong to triple-junction cells, at 32% at beginning of life and a projected 30% after 15 years' space exposure [79]. Figure 1.16 shows the assembly of current solar panels for use in space.

1.5 Summary

This chapter has described how photovoltaics evolved from some very simple experiments in the early nineteenth century to become a key enabling technology for the reliable functioning of satellite systems. Much of the current digital age relies heavily on the ability of satellites to communicate data; without them, modern society would

Figure 1.16 Assembly of solar arrays for use in space (Courtesy Spectrolab Corp)

not function. The development of solar cells for space use was the foundation stone for the large-scale terrestrial industry that exists today. Subsequent chapters in this book describe how this technology migration occurred and how it continues to develop toward large-scale deployment at the terawatt scale.

References

1 E. Roston: Bloomberg News, 4 November 2015.
2 Global Market Outlook 2019–2002 pub: Solar Power Europe (2019).
3 IEA PVPS Trends 2019.
4 DNV GL Energy Transition Outlook (ETO 2018).
5 W. Palz: 'Solar Power for the World' pub: Pan Stanford (2014).
6 J. Perlin: 'From Space to Earth – The Story of Solar Electricity' pub: Harvard University Press (2002) 1–13.
7 P. Varadi: 'Sun Above the Horizon' pub: Pan Stanford (2014).
8 J. Tsao, N. Lewis, and G. Crabtree: Solar FAQs, Sandia National Laboratory (2006).
9 J. Perlin: 'Let It Shine: The 6000 Year Story of Solar Energy' pub: New World Library (2014).
10 E. Becquerel: Les Comptes Rendus de l'Academie des Sciences 9 (1839) 561–567.
11 A. Smee: 'Elements of Electro-biology, or the Voltaic Mechanism of Man of Electropathology, Especially of the Nervous System and of Electro-therapeutics' pub: Longman, Brown, Green and Longmans (1849) 15.

12 W. Smith: Letter to Latimer Clark, Wharf Rd, 4 February 1873.

13 W. Smith: Journal of the Society of Telegraph Engineers, 2 (1873) 32.

14 W.G. Adams and R.E. Day: Philosophical Transactions of the Royal Society A 168 (1877) 341–342.

15 J. Perlin: 'From Space to Earth – The Story of Solar Electricity' pub: Harvard University Press (2002) 15–23.

16 J. Perlin: 'Let It Shine: The 6000 Year History of Solar Energy' pub: New World Library (2013) 306.

17 H. Hertz: Annalen der Physik 267, 8 (1887) 983–1000.

18 https://en.wikipedia.org/wiki/Bell_Labs accessed 7 August 2020.

19 M. Riorden and L Hoddeson: IEEE Spectrum 34, 6 (1997) 46–51.

20 R.S. Ohl: US Patent Application filed 27 May 1941.

21 https://en.wikipedia.org/wiki/History_of_the_transistor accessed 7 August 2020.

22 J. Perlin: 'From Space to Earth – The Story of Solar Electricity' pub: Harvard University Press (2002) 25–34.

23 C.S. Fuller: US Patent 3015590, 2 January 1962 (filed 5 March 1954).

24 D.M. Chapin, C.S. Fuller, and G.L. Pearson: US Patent 2780765, 5 February 1957 (filed 5 March 1954).

25 D.M. Chapin, C.S. Fuller, and G.L. Pearson: Journal of Applied Physics 25 (1954) 676.

26 Bell Laboratories Record November 1954 436.

27 Bell Laboratories Record May 1955 166.

28 M.A. Green et al.: Progress in Photovoltaics – Research and Applications 25 (2017) 3–13.

29 New York Times 26 April 1954.

30 US News & World Report 36 (1954) 18.

31 J. Perlin: 'From Space to Earth – The Story of Solar Electricity' pub: Harvard University Press (2002) 36–38.

32 H. Manchester: Readers Digest June (1957) 73.

33 M. Wolf: Proceedings of the 25th Power Sources Symposium (1972) 120.

34 J. Perlin: 'From Space to Earth – The Story of Solar Electricity' pub: Harvard University Press (2002) 41–48.

35 C.M. Green and M. Lomask: 'Vanguard, A History' NASA Document SP 4202 (1970).

36 NSSDCA/COSPAR ID: 1958-002B (1958).

37 J. Perlin: 'From Space to Earth – The Story of Solar Electricity' pub: Harvard University Press (2002) 49.

38 http://www.zarya.info/Diaries/Sputnik/Sputnik3.php accessed 7 August 2020.

39 J. Loferski: Progress in Photovoltaics – Research and Applications 1 (1993) 67–78.

40 J. Mandelkorn et al.: Proceedings of the 14th Annual Power Sources Conference (1960) 42.

41 W.G. Pfann and W. van Roosbroek: Journal of Applied Physics 25 (1954) 1422.

42 M.B. Prince: Journal of Applied Physics 26 (1955) 89.

43 E.D. Jackson: Transactions of the Conference on the use of Solar Energy, University of Arizona, Tucson, AZ, 5 (1955) 122.

44 W. Schockley and H.J. Quiesser: Journal of Applied Physics 32 (1961) 510.

45 NSSDCA/COSPAR ID: 1959-004A.

46 K.D. Smith et al.: Bell System Technical Journal 41(1963) 1765–1816.

47 M. Wolf: Solar Energy 5 (1961) 83–94.

48 J. Mandelkorn et al.: Journal of the Electrochemical Society 109 (1962) 313–318.

49 M.A. Green: Progress in Photovoltaics – Research and Applications 17 (2009) 183–189.

50 J. Lindmayer and J. Allison: Proceedings of the 9th IEEE PVSC (1972) 83–84.

51 J.F. Allison, R.A. Arndt, and A. Meulenberg: COMSAT Technical Review 5 (1975) 211–223.

52 J. Haynos et al.: International Conference on Photovoltaic Power Generation, Hamburg (1974) 487.

53 P.A. Iles: Proceedings of the 8th IEEE PVSC (1970) 345–352.

54 H. Fischer, E. Link, and W. Pschunder: Proceedings of the 8th IEEE PVSC (1970) 70–77.

55 J. Perlin: 'From Space to Earth – The Story of Solar Electricity' pub: Harvard University Press (2002) 50.

56 http://www.spectrolab.com/company.html#history accessed 7 August 2020.

57 G.P. Willeke and A. Räuber: Semiconductors and Semimetals 87 (2012) 7–48.

58 https://global.sharp/solar/en/ accessed 7 August 2020.

59 J. Perlin: 'From Space to Earth – The Story of Solar Electricity' pub: Harvard University Press (2002) 67.

60 J. Perlin: 'From Space to Earth – The Story of Solar Electricity' pub: Harvard University Press (2002) 117.

61 P.F. Varadi: 'Sun above the Horizon' pub: Pan Stanford (2014).

62 H. Cotal et al.: Energy and Environmental Science 2 (2009) 174–192.

63 P.A. Iles: Progress in Photovoltaics – Research and Applications 8 (2000) 39–51.

64 F. Dimroth: 'Photovoltaic Solar Energy: From Fundamentals to Applications' pub: Wiley (2017) 373–382.

65 V.M. Andreev: 'GaAs and High Efficiency Space Cells' pub: Elsevier (2003).

66 D.A. Jenny, J. Loferski, and P Rappaport: Physical Review 101 (1956) 1208.

67 J.C.C. Fan, C.O. Bozler, and R.W. McClelland: Proceedings of the 15th IEEE PVSC (1981) 672.

68 S.R. Kurtz, J.M. Olsen, and A. Kibbler: Proceedings of the 21st IEEE PVSC (1990) 138–140.

69 G.J. Bauhuis et al.: Solar Energy Materials and Solar Cells 93 (2008) 1488–1491.

70 https://www.businesswire.com/news/home/20180702005242/en/Alta-Devices-Breaks-Solar-Energy-Efficiency-Record accessed 7 August 2020.

71 Z.I. Alferov et al.: Soviet Physics Semiconductors 4 (1970) 2378–2379.

72 S.M. Bedair et al.: Proceedings of the 15th IEEE PVSC (1981) 21–26.

73 N.J. Ekins-Daukes in 'Solar Cell Materials – Developing Technologies' eds G.J. Conibeer and A. Willoughby pub: Wiley (2014) 113–143.

74 S.R. Kurtz and D.J. Friedman: National Centre for Photovolatics Program Review Meeting (1998) NREL/CP-520-25410.

75 Y.C.M. Yeh et al.: Proceedings of the 22nd IEEE PVSC (1991) 79–81.

76 Y.C.M. Yeh et al.: Proceedings of the 25th IEEE PVSC (1996) 187–190.

77 P.K. Chiang et al.: Proceedings of the 25th IEEE PVSC (1996) 183–186.

78 H. Yoon et al.: Progress in Photovoltaics – Research and Applications 13 (2005) 133–139.

79 http://www.spectrolab.com/photovoltaics.html accessed 7 August 2020.

2

The Beginnings of a Terrestrial Photovoltaics Industry

2.1   Introduction

At the end of the 1960s, silicon solar cells were unchallenged as the technology of choice for powering satellites in earth orbit and beyond. By 1970, however, there were concerns that despite intensive research in the previous decade, photovoltaic technology had plateaued on cost and performance and that it would not be able to supply power above 1 kWp for satellites [1]. A terrestrial market was virtually nonexistent, although Sharp Corporation in Japan had been active through the 1960s in using reject and surplus cells from satellites to power a range of off-grid applications, notably in coastal navigation [2]. Finding terrestrial applications had been a talking point in the United States. An early 1967 paper (see 3.1) looked at the potential for terrestrial photovoltaic systems. Its analysis gave the current cost of a photovoltaic system as $50/W and an estimated electricity cost of $2/kWh. The conclusion was unsurprisingly that photovoltaics would only be competitive in off-grid applications such as water pumping and lighting and that major cost reduction was needed. The 1970 IEEE PVSC conference had only 90 attendees, but nevertheless advocates like William Cherry maintained that solar energy had to be utilised to overcome long-term issues in fossil fuel supply [3], while Eugene Ralph asserted that with technology changes and scale-up to mass manufacturing, a factor 10 reduction in cost would occur, making terrestrial application viable [4]. Following this, in 1971, the National Science Foundation (NSF) under the aegis of the Research Applied to National Needs (RANN) programme began to seriously assess photovoltaics as a potential terrestrial energy source [5]. The joint NSF/NASA Solar Energy Panel met to set priorities and the NSF funded eleven projects at US universities with the goals of reducing cost from $50/Wp down to $5/Wp by 1977 and of offering a proof of concept for $0.5/Wp by 1979 and for 200 MWp installed by 1990. In retrospect, these goals seem wildly ambitious, but key research topics were correctly identified. While the technical potential was being recognised, however, there was no market pull to make these early concepts a reality – but then, in 1973, the oil crisis produced a major rethink in global energy supply. This chapter

Photovoltaics from Milliwatts to Gigawatts: Understanding Market and Technology Drivers toward Terawatts, First Edition. Tim Bruton.
© 2021 John Wiley & Sons Ltd. Published 2021 by John Wiley & Sons Ltd.

looks at how this crisis stimulated the development of terrestrial photovoltaic applications and how the photovoltaic technology evolved toward lower-cost mass production with the establishment of terrestrial-specific manufacturers.

2.2 The 1973 Oil Crisis

While there was a background belief amongst photovoltaic experts, it required a major international crisis and seismic shock in the world energy market to create mass recognition of the technology's potential. That crisis was the Arab–Israeli War of October 1973 and the accompanying oil embargo of the Organization of Arab Petroleum Exporting Countries (OAPEC) to Western nations – principally the United States, because of its ongoing political support of Israel and attendant large-scale supply of military hardware. Up to that point, the United States had enjoyed stable oil prices for decades, as shown in Figure 2.1, with oil being its major energy source for transport and a significant contributor to its electricity generation. However, as Figure 2.2 shows, domestic oil production had peaked in 1970, and imports were rising dramatically. The immediate consequences were a dramatic rise in fuel prices (the cost of petrol doubled) and petroleum shortages, with queues and violent clashes at filling stations.

The oil crisis wasn't driven simply by political concerns. As a result of the US government's allowing the dollar to float as a currency, oil revenues in the producing states had declined in real terms. Additionally, one of the key indicators in the oil industry – the ratio of proven new oil reserves to oil reserve depletion – fell below 1.0 (the ideal) in 1973, giving rise to the alarmist belief that the world would run out of oil in the 1990s.

In view of these pressures on the energy supply, President Richard Nixon announced in November 1974 'Project Independence' to make the country free from the need to import fossil fuel by 1980 [6]. The reality was that it took time for these

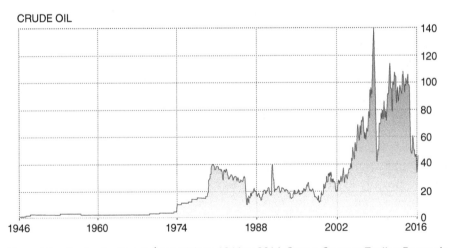

Figure 2.1   Crude oil prices in $/barrel from 1946 to 2016. *Source:* Courtesy Trading Economics

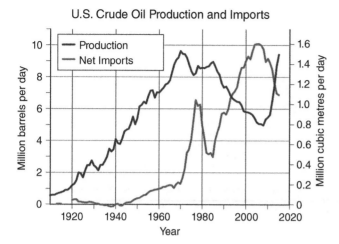

Figure 2.2 Import and export of crude oil in the United States from 1920 *Source:* US Energy Information Administration

changes to happen, and as shown in Figure 2.2 oil imports continued to rise strongly through the 1970s, leading to a second oil crisis in 1978, as can be seen from the price rise in Figure 2.1. However, agencies and programmes were established – notably the Energy Research and Demonstration Administration – which were effective in demonstrating significant advances in the longer term.

The real acceleration of the photovoltaic programme came under President Carter. His message to Congress in 1979 [7] looked forward to a time when the United States' major sources of energy would be derived from the sun. Interestingly, the motivation for this was freedom from future oil price shocks and embargos; climate change was not an issue at that time. Nonetheless, the support for renewables increased, and the influential Solar Energy Research Institute (SERI, later the National Renewable Energy Laboratory (NREL)) was finally opened in 1978, having being first announced in 1974. Figure 2.3 shows the US federal investment in energy research between 1961 and 2008 [8]. The peak in investment in all energy technologies is evident between 1974 and 1982.

When President Reagan took office in 1982, the immediate crisis in the energy sector was perceived to be over, and federal programmes were significantly cut back. For example, the budget at SERI was cut by 90%. Nevertheless, sufficient momentum had been generated in this period, and a real terrestrial photovoltaic industry had emerged which would continue to grow strongly, albeit with a much lower level of government funding.

2.3 The Way Ahead for Terrestrial Photovoltaics Technology

The oil embargo focussed efforts on the need to find new energy sources. An urgent conference was convened jointly by the NSF, the Jet Propulsion Laboratory (the NASA facility responsible for photovoltaic power), and the California Institute of Technology,

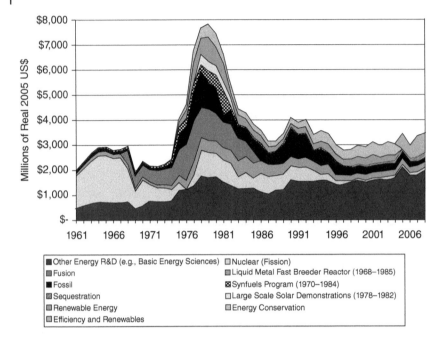

Figure 2.3 US federal investments in energy R&D 1961–2008 *Source:* J.J. Dooley: U.S Federal Investments in Energy R&D 1961–2008; US Department of Energy PNNL-17953, October 2008.

held at Cherry Hill, New Jersey on 23–25 October 1973 [9]. The meeting was attended by all the key photovoltaic scientists in the United States at the time, with the aim of establishing a consensus on plans, goals, and budgets and identifying the means for coordinating research, manufacturing, and commercial activities. Typical of the papers presented was Eugene Ralph's roadmap, summarised in Table 2.1.

Table 2.1 Roadmap for the achievement of low-cost solar systems [9]

	Milestone schedule		
	Technology status		
Parameter	1973 (present)	1978 (+5 years)	1983 (+10 years)
Cell area (cm^2)	20	45 or ribbon	Continuous sheet
Cell efficiency AM1 (%)	16.5	19	21
Cell cost ($/Wp AM1)	5	2.5	0.3
Power system cost ($/Wp AM1)	20	5	<1
Production rate (MWp/p.a.)	.09	6	200

Source: Workshop Proceedings, Photovoltaic Conversion of Solar Energy for Terrestrial Applications, Working Group and Panel Reports NSF-RA-N-74-013. Cherry Hill 23–25th Oct 1973

This gives an interesting view of the status of silicon technology in 1973. While 16.5% efficiency was probably the best case for small-area cells, 10% was more typical. The timescale for cost-effective development was again hugely underestimated, but the scenario for achieving the cost goals was not successfully followed until modern manufacturing lines with annual outputs above 1 GWp became operational after 2010. In these modern plants, cell efficiency is around 20% on large-area substrates. It should be noted that in 1973 the standard Czochralski (CZ) boule was 50 mm in diameter. The conference also included reviews of silicon-sheet technologies, as well as thin films and III–V devices and concentrating systems. Its conclusions were positive about the potential for photovoltaics to become a significant energy-generating technology within 10 years. It recommended a 10-year programme of $250 million for single-crystal silicon development and a further $45 million for realisation of multicrystalline silicon. The key target was the achievement of 10%-efficient solar cell with a module cost of $0.5/Wp (1975 dollars) and a lifetime of 20 years. The report of this conference and others led to the ERDA funding what was initially known as the Low Cost Silicon Solar Array (LSSA) project and later became the Flat Plate Solar Array (FPSA) project. This project was managed by NASA's Jet Propulsion Laboratory (JPL) from its start in 1975 to its close in 1986. Many of its developments were to transform the photovoltaics technology from a specialist space application to a viable terrestrial energy one. However, this wasn't the only photovoltaic activity: the concentrator development was established at Sandia Laboratories and thin-film research activities were parented at SERI.

2.3.1 Basic Silicon Photovoltaic Manufacturing Process

The manufacturing process for silicon photovoltaic technology is explained in detail in Chapter 4, but a schematic of the main process will be helpful in understanding the role played by the LSSA/FPSA and so is presented in Figure 2.4. This sequence is relevant to both space and terrestrial solar cells, with the main difference being in the final assembly. The feedstock for solar cells is semiconductor-grade polysilicon, with impurities in the part per billion level. In the early 1970s, only single-crystal silicon was used. It was normally ground to form a uniform cylinder, which was then sliced into wafers with a diamond saw. Early terrestrial solar cells were circular, but later the rods were shaped to have flat sides with curved corners, and pseudo-square wafers were processed to give better packing in the final module. The wafers were then cleaned and etched to remove saw damage and put through the solar cell fabrication process. The completed cells were soldered together using a tinned copper strip and encapsulated with a UV-stable polymer material with a low-iron toughened glass facing the sun and a polymer back sheet resistant to water ingress through the rear.

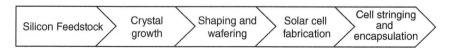

Figure 2.4 Schematic manufacturing process for silicon photovoltaic modules

2.3.2 The Flat Plate Solar Array Project

The FPSA project was initially set up to cover five technology areas, as described in this section [10]. The JPL programme was a very interactive one, with the aim not only of developing a technology but also of stimulating an industry to implement it. The work was organised through a series of contracts with universities, research centres, and industrial entities. In total, 256 contracts were let with 103 US institutions. JPL claimed that at some point every part of the US photovoltaics community had interacted with the project. Photovoltaics had proven its capability in space; the challenge for terrestrial applications was to make it affordable.

2.3.2.1 Solar-Grade Silicon

The aim of the programme was to develop a low-cost but high-purity silicon feedstock for less than $10/kg, as compared to the industry norm of $50/kg. The main contractor responsible for this development was the Union Carbide Corporation. Union Carbide successfully developed a silane-based chemistry to replace the conventional chlorosilane processes. A fluidised bed reactor was used to give a lower-cost product than the traditional Siemens process. In line with the programme goals, a 100 MT/year reactor was demonstrated and a 1200 MT/year commercial reactor was built, although its product was shipped not just to photovoltaics manufacturers but to the semiconductor industry in general. The Union Carbide plant was taken over by the Applied Silicon Materials Corporation but subsequently sold to a solar cell producer, the Renewable Energy Corporation (albeit a Norwegian-owned entity), so that the original programme goal of having a dedicated solar silicon feedstock supplier was ultimately achieved.

2.3.2.2 Silicon-Sheet Wafers and Ribbons

The price of CZ silicon wafer in 1975 was estimated at $10–12/Wp, which was clearly a barrier to achieving low costs. Rather than going through the extensive process of growing single crystals, grinding them, and sawing them into wafers, which entailed discarding as waste more than 50% of the high-cost starting material, it was decided to explore the possibility of going straight from feedstock to wafer. Two techniques were substantially supported: the dendritic web process, championed by the Westinghouse Corporation, and the edge defined growth (EDG) process, initially invented by Tyco Corporation, later transferred to Mobil Solar Corporation and then to the German ASE GmbH, and ultimately acquired by Schott Solar. Neither technique was ultimately successful. This subject will be discussed in more detail in Chapter 7.

2.3.2.3 High-Efficiency Solar Cells

In the latter part of the FPSA, it became apparent that increasing solar cell efficiency was an important element of cost reduction. From 1982 onward, research efforts were directed to this end, and the FPSA demonstrated $4\,cm^2$ silicon solar cells with efficiency up to 20.1% – one of the highest recorded at that time.

2.3.2.4 Process Development

A series of actions were sponsored for the improvement of solar cell processing, with good progress made in the areas of surface preparation, junction formation, and cell metallisation, as well as module assembly. Many processes were adopted by the manufacturing industry. For example, by 1977, Solarex was demonstrating ultrathin cells with antireflection coating and back surface fields.

2.3.2.5 Engineering Sciences and Reliability

Perhaps the greatest achievements came in the area of module reliability. At the start of the programme, modules were typically constructed of silicon solar cells manually soldered into strings, mounted on to a fibreglass board, and encapsulated in silicone rubber. The durability of these devices was poor, with lifetimes between six months and three years. Incidentally, this gave rise to the claim – which persisted for many years after module reliability had been improved enormously – that a silicon solar panel took more energy to manufacture that it generated in its lifetime. To improve durability, JPL introduced a series of block module purchases. A specification was produced for each block purchase in terms of the ability of the panel to resist certain accelerated lifetime tests, such as thermal cycling, humidity freeze, mechanical load, and hail impact. Block I was based essentially on the experience of JPL in qualifying space missions. The modules were field tested in challenging environments and their durability was assessed against laboratory stress testing. After each block, the severity of the tests was increased, until ultimately in Block V the environmental tests gave a good expectation of a 20 year module working life. A summary of the requirements of each stage is presented in Table 2.2, while Table 2.3 shows how the module structure

Table 2.2 Development of the FPSA block tests [11]

Test	I	II	III	IV	V
Thermal cycles	100–40 to 90 °C	50–40 to 90 °C	50–40 to 90 °C	50–40 to 90 °C	200–40 to 90 °C
Humidity	70 °C, 90%68 hours	5 –40 to 23 °C, 90%	5–40 to 23 °C, 90%	5–54 to 23 °C, 90%	10–85 to −40 °C, 85%
Hot spot (intrusive)					3 cells1000 hours
Mechanical load		100 cycles± 2400 Pa	100 cycles± 2400 Pa	10 000 cycles±2400 Pa	10 000 cycles±2400 Pa
Hail				9 impacts¾ in. 45 mph	10 impacts1 in. 52 mph
High potential		<15 µA1500 V	<50 µA1500 V	<50 µA1500 V	<50 µA2V system+1500 V

Source: J Wohlgemuth; "History of IEC Qualification Standards" International Module QA Forum 15th July 2011 NREL/PR-5200-52246

Table 2.3 Module procurement and structures in the FPSA block tests [12]

Block	I	II	III	IV	V
Date	Jun 1975	Dec 1976	May 1977	Nov 1978	Feb 1981
Area (m²)	0.1	0.4	0.3	0.6	1.1
Top cover	Silicone	Silicone	Silicone	Glass	Glass
Back plane	Rigid pan	Rigid pan	Rigid pan	Flexible sheet	Flexible laminate
Encapsulant	Silicone	Silicone	Silicone	PVB	EVA
Encapsulant method	Cast	Cast	Cast	Laminated	Laminated
Frame	No	Yes	Yes	Yes	No
Connections	Terminals	J box	Terminals	Pig-tails	Plug-in
Bypass diode	No	No	No	Yes	Yes
No. of cells	21	42	43	75	117
Cell size (mm)	76	76	76	95×95	100×100
Cell shape	Round	Round	Round	Shaped	Shaped
Silicon type	CZ	CZ	CZ	CZ	CZ
Junction structure	n/p	n/p	n/p	n/pp^+	n/p
Packing factor	0.54	0.6	0.65	0.78	0.89
NOCT (°C)	43	44	48	48	48
Module power (Wp) AM1.5, 28 °C	8	24	26	54	112
Module efficiency (%)	5.8	6.7	7.4	9.1	10.6
Encapsulated cell efficiency (%)	10.6	112	11.5	11.8	12.3

Source: M.I. Smokler et al; Proc 18th IEEE PVSC (1985) 1150-1158

changed over time from a very primitive design in Block I to something close to the modern module by Block V [12].

It was not until 1978 that the present module format arrived, with a toughened glass superstrate, polymer encapsulation by lamination, and a flexible backing sheet, initially Tedlar and ultimately a Tedlar/polyester/Tedlar or Tedlar/aluminium/Tedlar mix. Tedlar is a UV stable polyvinylfluoride polymer, developed in the 1960s by the Dupont Corporation. The development of the encapsulating polymers polyvinyl butylate (PVB) and ethyl vinyl acetate (EVA) is described in the next subsection. The cells were all single-crystal Czochralski silicon, initially round but by Block IV shaped with some flats for better packing. The first multicrystalline cells were also tested in Block IV. A Block VI was proposed but not implemented due to budget cuts.

In parallel with the JPL development, the ESTI at Ispra, Italy also developed a series of environmental tests (R501 and R502) in support of the European photovoltaics demonstration programme. These were similar to the Block V test, but with the addition of a UV exposure test and an outdoor exposure test and a reduction in the thermal cycling temperature from 90 to 85 °C.

The Block V specification and European specifications became the basis of the international standard IEC 61215, approved in 1991, which remains the basic requirement for all future silicon solar modules.

2.3.2.6 Module Encapsulation

The development of environmental stress testing demonstrated the need for improvements in module design and materials. For example, hail impact showed that toughened glass superstrates were required, while the silicone encapsulants were easily soiled and delaminated from the substrates. The challenge was to find a new encapsulation polymer. Some early modules had been made with PVB, which proved unstable, difficult to use, and expensive. Springborn Laboratories was successful in developing the EVA copolymer, which was crosslinked by thermal treatment up to 150 °C. Development began in 1975 with FPSA funding, and commercial production got underway in 1979 with Springborn's (later renamed STR) PhotoCap for use by the photovoltaics industry. The advantage of EVA was that it permitted dry lamination, which facilitated high-throughput and lower-cost module manufacture. Its impact is shown in Figure 2.5 [10], which charts how module lifetime improved over the course of the FPSA project.

In addition, a low-cost back sheet was required, as already described. Interaction between JPL and Dupont led to the adaptation of the highly UV-stable polymer Tedlar for photovoltaic module application.

2.3.2.7 Cost Goals

The overall aim of the FPSA programme was to make terrestrial photovoltaics affordable. The Cherry Hill target was a module cost of $0.5/Wp (1975 dollars) by 1985. Adjusting for inflation between 1973 and 1985, this became $1.07/Wp. Although this goal was not met by 1985, $1/Wp became the benchmark for the ultimate success of any photovoltaic technology going forward. Other studies, particularly by the EPRI, showed that a minimum module efficiency of 10% was necessary in order to keep the

Figure 2.5 Module lifetime over the duration of the FPSA programme [10] *Source:* W. Callaghan et al; Flat Plate Solar Array Project Final Report Oct 1986 JPL Publication 86-31

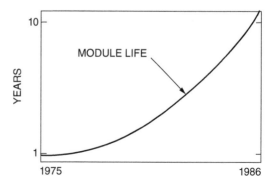

balance of system costs to reasonable levels [13]. The FPSA programme was able to track real market prices through its block purchases. In 1976, modules were bought from five manufacturers at an average price of $43/Wp (1985 dollars). By 1979, this had fallen to $20/Wp, and by 1983, large commercial projects were purchased at $5.65/Wp. Studies in 1985 predicted that a 25 MWp p.a. production unit could deliver product at $1.45/Wp. It should be noted that 25 MWp was about the size of the total global market at that time. Nevertheless, the FPSA programme had demonstrated a remarkable drop in prices, and if it was not quite achieved by 1985, the $1/Wp cost *was* ultimately achievable. It is also praiseworthy that the FPSA delivered its targets substantially with a total budget of $148 million (1974 dollars), as compared to the $295 million (1975 dollars) foreseen by the Cherry Hill meeting.

2.4   Rise of the US Photovoltaic Manufacturing Industry

While the government funding from 1974 onward was responsible for the dramatic strides forward as just described, the terrestrial manufacturing industry came about through the inspired actions of four entrepreneurs [14]. The first was Eliot Berman, who came not from a space cell background but from a photographics one. Berman joined Exxon in 1969 to research low-cost solar cells with potential to supply power to off-grid applications. His Solar Power Corporation was established in the early 1970s with a cost-reduction strategy of buying in off-specification wafers from the semiconductor industry. It went on to pioneer many of the applications for off-grid solar applications. Regrettably, despite its groundbreaking entry into the industry, Exxon was also one of the first companies to depart, closing down Solar Power Corporation in 1984. Next, from the space cell industry, came Bill Yerkes, who left Spectrolab to set up his own company, Solar Technology International, in 1975. That company sat at the forefront of cost reduction, with its use of screen printing of silicon contacts. It was sold to ARCO in 1978 to form ARCO Solar, which became the world's largest photovoltaics manufacturer by 1980 and had strong interests in the newer photovoltaic technology, setting up a joint venture with ECD to exploit amorphous silicon thin film and initiating its own programme in copper indium selenide. Third, Joseph Lindmayer was a semiconductor physicist who in 1973 was working with the satellite consortium COMSAT. He and the fourth entrepreneur, Peter Varadi, set up Solarex with a very limited budget but a strong drive to innovate new approaches. As well as developing cell technology, Solarex was at the forefront of the development of cast multicrystalline silicon for use in solar cells and, through its Swiss partner Intersemix, wire sawing of silicon wafers, in place of the expensive internal diamond-blade saws. Famously, in 1983, Solarex set up what it claimed would be a 'solar breeder factory', where a solar array on the south face of the building would generate the power needed to make more solar cells (Figure 2.6). Silicon casting, wafering, solar cell manufacturing, and module assembly were all carried out in the building, but it never became self-sufficient in

Figure 2.6   Solarex's 'solar breeder factory', Frederick, MD (Picture courtesy BP Archive)

electrical power. Solarex was bought by Amoco in 1983 and became part of BP Solar in 2000. Other major US companies involved in photovoltaics at the time were Westinghouse, which developed the dendritic web process, and Mobil Solar, which pioneered EDG. These were basically R&D activities with limited commercial production for field testing; they are described in some detail in Chapter 7.

2.5   Developments in Europe

Photovoltaics was slower to take off in Europe, partly because of the less strong technological base in space technology and partly because of the number of independent countries there. Nevertheless, there were similarities to the United States. The 1973 oil crisis did stimulate interest in renewables, and the UNESCO Solar Summit held in Paris in that year provided the same function as the Cherry Hill meeting. The response varied greatly by country. In the United Kingdom, photovoltaics was rejected and much greater priority was given to wind power. In France, there was an initial positive reaction and research activities were started, but these were later largely abandoned as the country chose the nuclear option [17]. Despite this, Philips, through its subsidiary at RTC in Caen, started photovoltaic development, leading eventually to the formation of the Photowatt company, which still exists today as part of the EDF utility company. Solarex also set up a number of factories and joint ventures in Europe, with France Photon in Angouleme, Holecsol in Eindhoven, Solare (with ENI) in Italy, and Intersemix in Switzerland [15].

Although Germany was to play a major role in the realisation of photovoltaics as a globally significant technology, its initial response to the 1973 oil crisis was slow [16].

The first Framework Programme Energy Research 1974–1977 did not include renewable energies, but things began to improve from 1977 onward with the programme Technologies for the Utilisation of Solar Energy 1977–1980. The AEG-Telfunken company had been making silicon space cells in Heilbron and undertaking system assembly in Wedel (Hamburg) from 1968. In the mid-1970s, it began to produce terrestrial modules. A particularly good collaboration was established with the Wacker company, which had previously developed a casting process for multicrystalline silicon: the SILSO process. In 1977, AEG-Telefunken started an eight-year collaboration with Wacker's solar subsidiary, Wacker Heliotronic GmbH. Wacker developed the casting process as well as a ribbon process and investigated upgrading metallurgical-grade silicon as a feedstock for the photovoltaic industry. The Siemens company also started its own R&D programme with the processing of single-crystal material, development of low-cost silicon feedstock, investigation of a silicon sheet process, and work on amorphous silicon. Nukem, a subsidiary of the major utility company RWE, looked at technology transfer for the Cu_2S/CdS process developed at the University of Stuttgart. In 1979, the aerospace company MBB also started an amorphous silicon research project. Key research institutes were set up which were to play important roles in the future development of photovoltaics. The Institute for Physical Electronics was established by Prof. Werner Bloss at the University of Stuttgart, while Prof. Adolf Goetzberger started the FhG.ISE in Freiburg. Prof. Bucher at the University of Konstanz led a significant group in silicon and thin-film technologies. More institutes would follow in the next decade. The greatest impulse came from the European Commission, which stimulated cross-border R&D from 1977 onward, when Dr Wolfgang Palz was recruited to lead its renewables programme [17]. This programme continues today, sponsoring major photovoltaic research projects. An important arm of the European programme was the demonstration activity, which saw many innovative applications of photovoltaics. For example, the Pellworm Island 200 kWp project was commissioned in 1983 to allow the island to have its own autonomous electricity supply (Figure 2.7) [18].

2.6 The Transition in Cell Technology from Space to Terrestrial Applications

Silicon solar cells for space had been intensively developed in the period from 1958 to 1973, as detailed in the 2nd to 18th IEEE PVSC Conferences. The requirements on the cells were different than for terrestrial applications [19]. They had to function in the AM0 spectrum, which is much bluer than the AM1.5 Global taken as the standard for the earth's surface, and under high levels of cosmic ray bombardment. A key parameter was their end-of-life efficiency following a nominal 10-year mission. They also had to withstand high levels of thermal shock, as coming out of the earth's shadow into sunlight could result in an almost instantaneous 140 °C temperature rise. A high level

Figure 2.7 Pellworm Island 200 kWp-array European demonstration project (Picture courtesy Hansewerk AG)

of reliability was required, too, as until the space shuttle was available, repair of satellites was impossible. Given the costs involved in building and launching a satellite, the cost of the individual solar cells was not a parameter of much concern; the critical number was the kWp/kg figure of merit. A major rethink was necessary in order to address how much lower-cost cells could be made for the terrestrial market.

The first issue was the silicon wafer. Space solar cells were generally rectangular – typically 2×1 cm, cut from a 100 m round silicon wafer. The rectangular shape allowed high packing density in the arrays deployed in space. The first cost saving came in processing the fully round wafer into a solar cell. Prime-grade wafers were used as supplied to the semiconductor industry. Bill Yerkes has described how at STI [20], they would purchase 1000 wafers from Wacker and process them into cells; once those were sold, they'd buy 1000 more. On taking over STI, ARCO thought it could obtain major cost reductions here, but Wacker saw no reason at the time to treat photovoltaic companies as a special case. The next stage in cost reduction came from using out-of-specification wafers or part-processed reject wafers from the semiconductor industry. SPC also set up CZ crystal growth using reject semiconductor-grade polysilicon feedstock and tops and tails from CZ growth from the major wafer producers [21]. This led to significant cost savings, although a 0.5 mm wafer thickness was typical at the time.

A typical space solar cell was of p type CZ silicon with a phosphorus-diffused emitter and evaporated titanium–palladium–silver contacts. The phosphorus diffusion wasn't particularly a cost issue: the existing tube diffusion furnaces using the $POCl_2$ liquid source were well established, and surplus equipment from the semiconductor industry provided a low-capital-investment equipment source. However, spin and spray-on phosphorus sources had been developed and were used in belt furnaces for a much more rapid throughput than that obtainable in the conventional tube furnace. The major cost issue was the titanium–palladium–silver contacts. These were formed using photolithography, which was an expensive process both in terms of the cost of the photolithographic chemical and in the fact that the high-value metals were deposited over the whole surface and then 90% of it was etched away and discarded. A lower-cost approach was thus needed. Two were pursued: plating and screen printing.

Plating had already been used for some space cells. Electroless nickel plating was well established for silicon. The nickel seed layer was sintered to form nickel silicide, giving an Ohmic contact and a diffusion barrier. The nickel layer could then be thickened with electroless copper and capped with a solder layer, or else solder could be applied directly to the nickel [22]. This approach was followed by Solavolt and SPC, but both ceased operations in the early 1980s.

The much more promising approach was screen printing, which could be used for both front and back contacts. This had the advantage that high-throughput machinery already existed for the hybrid circuit industry and the utilisation of the metal pastes was very high. One of the first reported uses of screen printing of solar cells was by E. Ralph at Spectrolab in 1975, when he recorded 10.7% (AM0 spectrum) solar cell efficiency for a 20.7 cm^2 solar cell made by a 'vacuum-free' process [23]. The process sequence was not too far different from that used in today's manufacturing, but at that time the silver pastes for the front contact had not been optimised. A key factor was the use of a screen-printed aluminium contact on the rear to give a good Ohmic contact to p type silicon. A titania/silica spin-on glass was used as the antireflection coating. The cell parameters are given in Table 2.4. The solar cell efficiency was measured at well below a normal 100 mW/cm^2, apparently so that series resistance would not limit the achievement of a >10% efficiency.

Clearly, interest in the screen printing technology started at this point. A further innovation was the ability to print the silver front contact on to the ARC, so that during the paste-firing process the paste etched through it to make ohmic contact, in what is now

Table 2.4 Solar cell parameters for a screen-printed silver front contact and aluminium rear contact [23]

Cell area	20.4 cm^2
Cell thickness	225 µm
AM0 insolation	14 mW/cm^2
Voc	~590 mV
Jsc	24.5 mA/cm^2 (corrected to 100 mW/cm^2)
FF	.70
Efficiency	10.7%

Source: E.L. Ralph; Proc 11th IEEE PVSC (1975) 315–316

known as 'firing through' the ARC. This method was reported as early as 1976 (3.5.8), when it successfully produced a 9.5% (AM1 spectrum)-efficiency solar cell on a 12.3 cm^2 semicircular silicon wafer. The work was done at the Ferranti company in the United Kingdom; Ferranti had been a supplier of space cells to the UK programme and had produced some terrestrial modules, but it had stopped trading by the 1980s.

The screen printing approach was pioneered in large-scale production by Yerkes on his setting up STI (he had previously worked at Spectrolab). The rise of photovoltaic manufacturing stimulated development of silver pastes for photovoltaics, in which the Ferro Corporation took a particular lead. By 1985, all commercial silicon solar cells were screen printed.

2.7 Alternatives to Silicon for Solar Cells

With the upsurge of interest in the use of photovoltaics for terrestrial applications, it was inevitable that the existing silicon cell technology in the early 1970s would be the starting point for the new solar industry. However, any semiconductor with a bandgap between 1.0 and 2.0 eV is potentially useable as a solar cell. In parallel with developments in silicon, other technologies were thus investigated.

At the same time as Bell Labs successfully demonstrated the silicon solar cell, the US Air Force Laboratories announced a working solar cell using a cadmium sulphide/copper sulphide heterojunction, although the efficiency was very low [24]. Silicon is an indirect bandgap semiconductor and therefore a relatively poor absorber of light, and so silicon solar cells must be more than 100 microns thick to give effective current generation without complex light-trapping structures. In contrast, a direct bandgap semiconductor can absorb the solar spectrum in 1 micron or less of absorber. When priority was given to cost reduction, it seems the obvious development path was to replace a 500 micron-thick silicon wafer with 1 micron of another semiconductor as a thin film. Thus thin film had the further attraction of being capable of being produced in large areas and patterned to give the high voltage necessary for terrestrial application, without the sawing, cell process, soldering, and stringing operations necessary for silicon module production. The motivation for the introduction of thin-film photovoltaics, as well as its evolution and current challenges, is described in Chapter 7.

In the 1970s, the foundations were put in place for later manufacturing operations. The first thin-film device to receive significant attention was the copper sulphide/cadmium sulphide solar cell. Following its initial discovery in 1954, significant improvement was made by Shirland, and in 1965 a 5.8% efficiency was reported on a 50 cm^2 solar cell [25]. This cell was made by vacuum evaporation of cadmium sulphide on to a molybdenum substrate followed by heat treatment. The substrate was then dipped into aqueous cupric chloride solution, where an ion-exchange reaction occurred and a p type cuprous sulphide was formed. A metal grid was applied to the Cu$_2$S layer and the device was encapsulated.

This research led to the production of cells by the Clevite Company, although these proved to be unstable on exposure to light [26]. (However, some cells were stable, and a working 50-year-old Cu_2S/CdS Clevite cell is held at the University of Delaware's Solar Museum [27].) Nevertheless, research effort continued, and by 1973 the NSF/RANN programme was funding work on Cu_2S cells at the Institute of Energy Conversion (IEC) at the University of Delaware and Brown University [5]. The work at the IEC made good progress under the leadership of Karl Boer, and by 1971 8%-efficient solar cells were produced. Solar Energy Systems Inc. was formed in 1973 and received funding from the Shell Oil Company to set up a pilot line [28]. Research was also carried on elsewhere, with active programmes on Cu_2S/CdS cells at RTC Caen and SAT, Paris prior to 1973. However, although work continued through the 1970s, it was apparent by 1981 that other thin-film technologies were showing more promise and that the instability issues were proving intractable.

While there was research on other thin-film technologies such as cadmium telluride and copper indium diselenide, hydrogenated amorphous silicon solar cells emerged in the late 1970s as the most promising thin-film technology. As early as 1969, Chittick et al. [29] had reported the deposition of hydrogenated amorphous silicon by RF glow discharge of silane and produced material with photoconductive properties similar to those of cadmium sulphide. This work was further developed by Spear and Lecomber at Dundee University (who demonstrated p and n doping in amorphous silicon [30]). The real breakthrough came from the RCA Laboratories in the United States in 1976, when Carlson and Wronski reported the first p-i-n amorphous silicon solar cell, with an efficiency of 2.4% in an AM1 spectrum [31]. This gave rise to a further three decades of intensive development of amorphous silicon. As described in Chapter 7, however, the amorphous silicon approach was not ultimately successful, but thin-film modules based on cadmium telluride and copper indium diselenide and its derivatives continue to be manufactured commercially.

2.8 Summary

This chapter has explained how solar cells transitioned from being a niche high-cost energy source for space satellites to becoming a growing industry supplying terrestrial applications. This move was driven by the oil crisis of 1973, when it became important to find sustainable alternatives to imported oil supplies as a prime energy source. While space cells functioned well, they were extremely costly, and the challenge was to make them more cost-effective by modifying the cell-fabrication process and introducing the concept of module formation to give a product capable of operation for twenty years across the range of earth's climate types and locations. Production facilities were established for silicon solar modules in a number of countries; the next chapter explains how the early market developed.

References

1 J. Loferski: 'Solar Power for the World' ed. W. Palz pub: Pan Stanford (2014) 281.
2 J. Perlin: 'From Space to Earth –The Story of Solar Electricity' pub: Harvard University Press (2002) 67.
3 W.R. Cherry: Proceedings of the 8th IEEE PVSC (1970) 331–337.
4 E.L. Ralph: Proceedings of the 8th IEEE PVSC (1970) 327–330.
5 R.H. Blieden and F.H. Morse: Proceedings of the 10th IEEE PVSC (1973) 216–226.
6 https://www.presidency.ucsb.edu/documents/address-the-nation-about-national-energy-policy accessed 7 August 2020.
7 https://www.presidency.ucsb.edu/documents/national-energy-program-fact-sheet-the-presidents-program accessed 7 August 2020.
8 J.J. Dooley: 'US Federal Investments in Energy R&D 1961–2008' US Department of Energy October 2008 PNNL-17953.
9 Workshop Proceedings, Photovoltaic Conversion of Solar Energy for Terrestrial Applications, Working Group and Panel Reports NSF-RA-N-74-013; Cherry Hill, NJ, 23–25 October 1973.
10 W. Callaghan et al.: 'Flat Plate Solar Array Project Final Report' JPL Publication (1986) 86–31.
11 J. Wohlgemuth: 'History of IEC Qualification Standards'; International Module QA Forum 15 July 2011; NREL/PR-5200-52246.
12 M.I. Smokler et al.: Proceedings of the 18th IEEE PVSC (1985) 1150–1158.
13 E.A. DeMeo, D.F. Spencer, and P.B. Bos: Proceedings of the 12th IEEE PVSC (1976) 653–657.
14 J. Perlin: 'From Space to Earth – The Story of Solar Electricity' pub: Harvard University Press (2002) 49–56.
15 P.F. Varadi: 'Sun Above the Horizon: The Meteoric Rise of the Solar Industry' pub: Pan Stanford (2014).
16 G.P. Willeke and A. Raeuber: Semiconductors and Semimetals, 87 (2012) 7–48.
17 W. Palz: 'Solar Power for the World' ed. W. Palz pub: Pan Stanford (2014) 68–70.
18 C. Kruse, H.J. Lowalt, and K. Maass: Proceedings of the 10th EUPVSEC (1991) 724–727.
19 P.A. Iles: Progress in Photovoltaics: Research and Applications, 8 (2000) 39–51.
20 J.W. Yerkes: private communication.
21 P. Caruso: interview with the author 1984.
22 M.G. Coleman, R.A. Pryor, and T.G. Sparks: Proceedings of the 14th IEEE PVSC (1980) 793–799.
23 E.L. Ralph: Proceedings of the 11th IEEE PVSC (1975) 315–316.
24 J. Perlin: 'From Space to Earth – The Story of Solar Electricity' pub: Harvard University Press (2002) 173.
25 F.A. Shirland and J.R. Hietenan: Proceedings of the 5th IEEE PVSC (1965) C3-1–C3-15.

26 J. Perlin: 'From Space to Earth – The Story of Solar Electricity' pub: Harvard University Press (2000) 49–56.

27 https://www.youtube.com/watch?v=DrHUrY2xNgU accessed 7 August 2020.

28 K.W. Boer: 'Solar Power for the World' ed. W. Palz pub: Pan Stanford (2014) 219–226.

29 R.C. Chittick et al.: Journal of the Electrochemical Society, 116 (1969) 77–81.

30 W.E. Spear and P.J. Lecomber: Solid State Communications, 17 (1975) 1193–1196.

31 D.E. Carlson and C.R. Wronski; Applied Physics Letters, 28 (1976) 671.

3

The Early Photovoltaic Global Market and Manufacturers

3.1 Introduction

The previous chapter described how photovoltaics came to be recognised as a technology with the potential to be a significant generator of electricity for the global energy supply. This chapter looks at how the first terrestrial applications were found and how the early photovoltaic market was established. While government programmes in the 1970s and early 1980s focussed on supporting R&D programmes to achieve the major cost reduction needed to open up the terrestrial market, there was very little thought given to how an actual market might evolve. One barrier was the fact that the targets for cost reduction were hopelessly optimistic, with 1987 seen as the year in which the ideal of $1/Wp module cost might be achieved [1]. One of the earliest papers from the NSF programme [2] suggested the market would be for self-consumption in buildings, for central power generation, and for beaming power to earth from large solar arrays in space. Some very creative ideas emerged. One, by the distinguished photovoltaic pioneer William Cherry, imagined a 1 square mile airborne solar mattress, helium-filled and tethered at between 50 000 and 70 000 feet altitude and capable of generating 250 MWp. It would be positioned in areas of high electricity peak demand to avoid brownouts [3]. While the solar mattress did not get built, self-consumption in buildings and central generation did develop over time. The concept of beaming power down from space has not been realised, although the concept still attracts interest [4]. A few embryonic manufacturers were in place in 1980, and while they received R&D support, the onus was on them to find their own markets and become financially viable.

Early work showed that the relatively high cost of photovoltaics meant only standalone systems of less than 5 kWp were viable and that battery storage was an essential component of any application [5]. Additionally, installations were often at sites where there was difficult access, so low maintenance was a requirement. It is easy to see how the first application of photovoltaics as a power supply for satellites was an extreme example of this. Fortunately, the appearance of photovoltaic companies in the 1970s was coincident with the emergence of new applications for which

Photovoltaics from Milliwatts to Gigawatts: Understanding Market and Technology Drivers toward Terawatts, First Edition. Tim Bruton.

Figure 3.1 Photovoltaic panels on an oil production platform (Courtesy BP Archive)

photovoltaic power was the most appropriate solution. For example, one of the early applications was supplying navigation warning lights for oil production rigs in the Gulf of Mexico. A typical unmanned off-shore platform with photovoltaic modules is shown in Figure 3.1. Prior to the 1970s, warning lights were powered by large and expensive nonrechargeable batteries [6]. The advantages of photovoltaic power were obvious, in that it didn't require the regular transfer of these heavy batteries to and from small boats. From 1978 onward, Tideland Signal and other companies started to supply solar-powered warning lights and other navigation aids [7].

The off-grid market in the 1980s developed into distinct segments. Initially, the largest was the off-grid professional market, which included navigation aids and telecommunications. Next was the off-grid domestic market, which included remote homes and communities with no grid connection, spreading into similarly remote schools and hospitals. A subset of the domestic market was the indoor consumer market, supplying such applications as watches, calculators, and radios. While most of the domestic off-grid market was in developing countries, the consumer market was in wealthier ones. Although not the main market, some early demonstrations of large central grid-connected plants did take place. The evolution of each of these sectors will be described in detail in the rest of this chapter.

3.2 Off-Grid Professional Market

3.2.1 Navigation Aids

This was the one of the earliest sectors to emerge, in the 1960s. Japan led the way, with the first demonstration of a solar-powered lighthouse with 225 Wp of solar modules in 1966 (Figure 3.2), and with 4.5 kWp of marine systems installed by 1967 – although only with cells not wanted for the space market [5,8]. The oil industry was a big

Figure 3.2 Ogami Island lighthouse, Nagasaki, 1966. *Source:* Japanese Cost Guard

customer in the Gulf of Mexico, but the greatest boost came from the US Coast Guard, which received funding from the Department of Energy in 1977 for a demonstration of solar lighting for navigation buoys [8]. Previously, the buoys had been powered by nonrechargeable batteries, which had to be taken out by boat, or by acetylene gas lamps, which presented major maintenance issues. The saving here was not in the absolute energy cost but in the removal of the need for frequent maintenance visits. However, it took some time for solar power to be accepted in the marine environment. One issue was that buoys are subject to frequent immersion in salt water. Pioneering work was carried out by Captain Lloyd R. Lomer of the US Coast Guard, who developed more rigorous environmental tests than the then current JPL Block V test, leading to the USCG PV Power Test Specification, which was issued in November 1981 [8].

Still, the fact that this was such a cost-effective application and that companies were now in a position to manufacture reliable photovoltaic panels meant that there was a rapid roll-out of solar-powered navigation aids by the early 1980s. In 1983, an acetylene-powered buoy sold for $15 300, while the photovoltaic alternative cost only $2000 [8]. Canada was quick to follow the United States' example. France began replacing its acetylene lighthouses with photovoltaic ones in 1981, while the majority of Greece's 960 buoys and lighthouses were photovoltaic-powered by 1983 [9]. As early as 1977, Germany had deployed photovoltaic-powered pollution-monitoring buoys in the River Elbe [9]. The United Kingdom also took up the challenge; currently, all of its lighted buoys are solar-powered, as are 30 of its lighthouses [10].

3.2.2 Microwave Repeater Stations

In the early 1970s, microwave repeaters for telecommunications were massive affairs with inefficient electronic loads requiring large amounts of power and good access for frequent maintenance. By the middle of the decade, a major breakthrough with much simpler electronics occurred, greatly reducing the power requirement to that which could be supplied by a few solar panels [11]. The first commercial photovoltaic system was installed at Hunts Mesa by Navajo Communications and was a great success – so much so that within six years one company, GTE, had sold 1000 solar-powered repeater stations. The low cost, high reliability, and low maintenance requirement made it cost-effective for the first time to supply smaller communities across the United States; previously, the residents of one small town – Cuprum, Idaho – had had to make a 55-minute car journey to place a long-distance telephone call.

The potential was quickly realised in other countries – most notably, Australia, which likewise had a very scattered rural population. The earliest systems consisted of a telephone connected to a radio transmitter, which would connect to a relay station connected to the national network. The major challenge came in powering the system. Thermoelectric generators, wind turbines, and diesels were all tried, but were found to be unreliable. Early experiments were conducted with high-cost Sharp modules, but by 1976 Solar Power Corp. modules were available at $20/Wp, and after successfully being

Figure 3.3 Hybrid photovoltaic–diesel microwave repeater station in Papua New Guinea, 1980s (*Source:* Courtesy BP Archive)

rolled out for single telephones, they were considered for use in repeater stations across the country. In 1978, one of the first applications of a large system was in connecting Tenant Creek to Alice Springs – a distance of 580 km – with 13 microwave repeater stations ranging from 132 to 143 W of continuous power [11]. By 1985, systems up to 1 kW continuous were common, while higher-power systems often used hybrid solar–diesel combinations. Figure 3.3 shows a hybrid photovoltaic/diesel repeater station in Papua New Guinea in the 1980s, where only access by helicopter is possible.

High system reliability was essential, in order to keep maintenance visits to a minimum. Improvements were necessary in deep-discharge lead acid batteries and charge control, and batteries were developed specifically for photovoltaic applications. Sizing programmes were designed to predict the battery size required for high reliability of operation in various climates. Generally, battery storage capacity increased with latitude, from 5 days in the tropics to 25 days at 50° N [12]. The early successes at Telecom Australia led to photovoltaic power being the technology of choice for repeater stations. A system was developed in which all components were put into a shipping container, which then housed the system electronics in operation. The Kimberley project in North West Australia, commissioned in 1983, remains the world's largest solar-powered microwave repeater network, spanning 2415 km with 43 repeaters [13].

Many other countries emulated the experience in Australia, particularly those with remote populations such as Colombia and Botswana [12]. France deployed a 1 kWp photovoltaic-powered microwave repeater station in 1977 [14], while Nepal used photovoltaic power for communications as early as 1978 [15]. This application has continued up to the present time, so that photovoltaics is the technology of choice for powering microwave repeaters today [16].

3.2.3 Cathodic Protection

Cathodic protection was first proposed by Sir Humphrey Davy in 1824 [17]. This involved providing a 'sacrificial' anode to be corroded in place of an important metal that one wanted to keep protected. It was later found that impressing current on a pipeline (impressed current cathodic protection, ICCP) had the same effect [18]. The ICCP concept was applied to pipelines in the United States in the 1930s, but the issue was how to supply the power over long distances through the remote areas where pipelines typically ran. The same problems experienced in the telecoms sector applied.

The solar age in ICCP began in the 1970s when problems were encountered with the corrosion of well heads in the Hugoton gas field, extending from South West Kansas through into Oklahoma and Texas. The issue was exacerbated as the pipeline ran through a salt marsh region. A first test with a module built with reject Spectrolab cells proved effective, and when lower-cost modules became freely available from Solar Power Corporation, the deployment rapidly took off [18]. A typical modern system is shown in Figure 3.4.

Following this early success, the application became well established by 1980. It is estimated that there are between 4000 and 5000 systems operational in the United States today, with many more elsewhere around the world.

Figure 3.4 Typical photovoltaic-powered ICCP system in the Libyan Oil Field. *Source:* Solapak Systems Ltd.

3.2.4 Other Applications

While the three applications just described are the largest segments of the off-grid professional market, there are a whole range of small-scale applications for which photovoltaics is an ideal source of electrical power. It was recognised as early as 1981 that photovoltaics was the most cost-effective option for loads less than 1 kW and possibly up to 5 kW if the equipment was more than 1.7 km from an overhead power line or 0.6 km from a buried one [14]. This opened up a whole host of potential applications, from radio beacons though to weather monitoring and emergency refuges in rural areas. Since the 1980s, the list has grown to seem endless.

The use of photovoltaics removes the need for expensive cabling and reduces installation times and costs. A common sight in the United Kingdom and elsewhere is photovoltaic-powered parking meters (Figure 3.5) and traffic speed measurement devices (Figure 3.6). These have the advantage that they can be easily moved to different locations as needed. Likewise, between 2006 and 2011, the UK Highways Agency

Figure 3.5 Photovoltaic-powered parking meter (Courtesy of Bruton)

Figure 3.6 Photovoltaic-powered vehicle speed measurement (Courtesy of Bruton)

Figure 3.7 UK roadside emergency telephone (*Source:* GAI-Tronics)

Figure 3.8 Solar-powered street light on a building site (*Source:* Prolectric Ltd)

replaced all 9500 of its roadside emergency telephones with solar-powered wireless versions (Figure 3.7). The advantages here are obvious, in that no cables need be supplied to the installation. A 1.7 Wp solar panel is adequate to power a telephone device even in the moderate solar insolation of the United Kingdom [19].

The provision of street lighting is another example where ease of installation is a factor, as is the reduced need for electrical power – a serious consideration in many developing countries. A typical example is shown in Figure 3.8. This is actually a modern version using LED lights. Its versatility is emphasised by the concrete base, which provides stability but also allows the light to be moved and repositioned with a forklift truck [20].

3.2.5 Early Grid-Connected Application

While the commercial market was dominated by off-grid applications, many had seen the future of photovoltaics as providing large grid-connected power plants to replace fossil-fuelled ones. For example, the US Electric Power Research Institute stated in the

Figure 3.9 Two-axis tracked mirror-enhanced solar revivers at the Carrisa Plains photovoltaic plant (*Source:* Fluor Corp.)

August 1983 edition of its journal that 'for solar cells to make a significant impact on electricity production in this country ... they will have to go for the real thing: bulk electric power generation'. This view was widely held, and there were a number of demonstrations of large grid-connected photovoltaic plants. The first 1 MWp plant in the world was built by ARCO Solar at Hesperia, California on Southern California Edison land close to a utility substation in November 1982 [22]. The field consisted of 108 two-axis trackers – established sun trackers developed for heliostat solar thermal applications – each of which supported 16 standard ARCO solar 1×4 ft monocrystalline modules. The power generated was delivered to the grid via a 12 kV cable at the substation.

As soon as the Hesperia plant began operations, ARCO Solar proceeded to build a 6 MWp plant at the Carrisa Plains on Pacific Gas and Electric land in California (Figure 3.9) [22]. This plant was ambitious not only in its scale but in the way it combined of two-axis tracking of the sun with planar mirror enhancement of the incoming light to the module. However, the installation was only a limited success and showed significant degradation in its power output. This was initially blamed on 'browning' of the EVA caused by the high ambient temperature and additional solar irradiation from the mirrors [23]. However, it was subsequently shown that poor soldering was a more significant contributor [24]. The plant was decommissioned in 1994 and the site is now derelict. It was perhaps 'a solar plant before its time' [25].

Browning of modules, particularly in hot climates, was a general problem in the 1980s, but was largely solved by improvements to the EVA formulation and by the recommendation to use cerium oxide in the low-iron module cover glass by the early 1990s [26].

This bad experience did not stop the enthusiasm for large central plants. The Sacramento Municipal Utility District (SMUD) installed a 1 MWp plant in 1984 [27]. In Japan, a megawatt-size plant was constructed at Seijo City in 1985. The first grid-connected plant in Europe was the 30 kWp one constructed at Marchwood in the United Kingdom in 1983, with funding from the European Commission (Figure 3.10) [28].

Figure 3.10 First European central grid-connected photovoltaic power plant (30 kWp) at Marchwood, UK (Courtesy BP Archive)

While the early centrally grid-connected plants were constructed as demonstrations, it was not at all obvious in the 1980s that this was the way forward for photovoltaics. It had been observed that electricity was used everywhere and that the sun also delivered energy in most locations, and so there should be no need for a high-voltage transmission line, which involved significant costs and losses [29]. The most striking initiative for decentralised grid connection was taken by the Apha Real Company in Switzerland in the mid-1980s; they promoted the idea of a 1 MWp system, but as 333 homes each with a 3 kWp array [30]. The idea was rapidly adopted by an enthusiastic Swiss population, and the 1 MWp of grid connected solar power was installed. Back in California, SMUD quickly switched from running large central plants to renting individual roofs. This became the chief application for photovoltaics and the motor for growth from the 1990s until well after 2010, as will be discussed later in this book.

3.3 Off-Grid Domestic Market

As described in the previous chapter, the focus of the early development in photovoltaics was in industrialised countries, with the objective of securing energy supply and freedom from embargoes and price shocks, as noted by US President Jimmy Carter in 1977 [31]. Concern about climate change and greenhouse gas emissions did not really become a topic justifying photovoltaics deployment until after the Earth Summit Conference in Rio de Janeiro in 1992. In the 1980s, the major concern was the recognition that 1.7 billion people in the world had no grid electricity supply and that there was little prospect of their gaining it in the medium term. This was viewed as both a business opportunity for the emerging photovoltaic companies and an ethical issue, as those without electricity were typically among the poorest people in the world [32]. In the late 1970s and through the 1990s, the off-grid market was the mainstay of the emerging photovoltaic industry. The main applications were lighting for homes and community centres, power for health clinics, water pumping, and communications. In addition, a significant sector emerged in developed countries around small electronic devices such as watches, calculators, and games, where solar power was simpler, cheaper, and more convenient than batteries.

3.3.1 Solar Home Systems

The provision of electricity to individual homes and small communities with no grid power was one of the first markets to emerge. A typical solar home system consisted of a photovoltaic module, a battery, a charge controller, up to three fluorescent lights, and a power socket capable of powering a black and white television for up to 3 hours. The photovoltaic system gave not only a more reliable power supply but also a better quality of light without the risk of fire inherent in kerosene lamps and candles (Figure 3.11).

The photovoltaic development programme in the United States funded a range of solar-powered buildings, including the world's first standalone solar-powered community, the Schuchuli Native American Community in Arizona, founded in 1978 [33]. At around the same time, to offset criticism of French atomic bomb testing in the Pacific, a programme of electrification began in the French overseas territories. While a number of alternatives were investigated, it quickly became apparent that the only viable solution for low-cost reliable small-scale electrification was photovoltaic power [34]. The programme was so successful that, by 1987, 3300 houses in French Polynesia were solar-powered – about 50% of all homes in the region that received electricity. In addition to government programmes, a growing private market developed in many countries in the late 1970s and early 1980s. As with navigation aids, prior to the availability of photovoltaic modules, many systems relied on transporting lead acid batteries by jeep over large distances, as Wolfgang Palz records for the supply of electricity to schools in French-speaking West Africa [35]. John Perlin records a similar experience in Kenya, where rural middle-class Kenyans could access electricity only by taking lead acid batteries some distance to a recharging facility, leading to the rapid adoption of photovoltaics [36]. A private market also arose in developed countries that had many remote homes or weekend cottages without any grid connection, such as Australia, Spain, Norway, and the United States (where Alaska was one of the largest early markets) [37].

While these various programmes provided solar home systems, a number of problems in local acceptance, maintenance, and sustainable funding emerged. NGOs began to recognise these problems and to put corrective action in place. For example, Neville Williams set up the Solar Electric Light Fund, which provided photovoltaic

Figure 3.11 Lighting in Africa: (a) photovoltaic-powered; (Courtesy BP Archive) (b) candle (Courtesy BP Archive)

(a)

(b)

lighting and sufficient power for a black and white television to communities in Sri Lanka, followed quickly by India, Nepal, China, Vietnam, and South Africa. The installations were financially self-sustaining, with local funding mechanisms, a high local content, and technical support [38].

From its humble beginnings, the solar home system became a well-established means of alleviating poverty and improving quality of life in many countries. The World Bank has spent $1.4 billion dollars on photovoltaic systems to date, benefiting 15 million people in 30 countries [39]. While it does not utilise a significant fraction of

Table 3.1 Impact of the installation of 1.5 million solar lights in Africa

Number of people served	10 million
Money saved	£340 million
People enjoying better health	5.8 million
Extra study hours	2 billion
CO_2 emission saved	88 0000 tonnes

Source: www. Solar-aid.org accessed 8/9/2016

the world's photovoltaic output, the social and environmental significance of this sector is very large. The impact of the deployment of 1.5 million solar lights in Africa can be seen in Table 3.1, for example [40].

3.3.2 Water Pumping

Access to good-quality water has been and remains a major problem for many communities around the world. Even today, 2 billion people do not have access to safe drinking water [41]. At the same time that solar home systems were being recognised as a means of rural electrification, photovoltaics was also coming to be seen as a potential source of power for water pumping. The same problems found in powering remote navigation aids and telecommunications were also seen in this application. Diesel power was unreliable in remote locations. Solar thermal pumps with steam generation were difficult to install and maintain. Hand pumps, while low-cost, gave relatively little water and required regular maintenance [42].

The first successful solar photovoltaic water pumping operation was carried out in Corsica in 1973 [43]. However, the real pioneer of the method was Father Bernard Verspieren, who, having seen many people dying of drought, set out to improve the water supply in remote locations in Mali [44]. Witnessing the photovoltaic pump in action in Corsica, he immediately recognised its potential for use in Africa. With voluntary donations, the first photovoltaic water pump was installed in Nabasso, Mali and worked without breakdown for 3 years. The project was rolled out across the country with Solar Power Corporation modules and Guinard (France) pumps. By 1981, Father Verspeiren had become the second largest private purchaser of photovoltaic panels in the world, after Telecom Australia. The potential for photovoltaic water pumping encouraged Guinard to set up a joint venture with Solarex, named France Photon, to manufacture photovoltaic panels [45].

A disadvantage of the Guinard pump was that the motor was located above ground, with a long rigid shaft down the well. In the early 1980s, the Danish manufacturer Grundfos made significant improvements with a low-maintenance down-the-borehole pump with flexible plastic tube to deliver water to the surface, which coped much better with the irregular bores drilled in Africa [46]. There was a rapid

Figure 3.12 Photovoltaic-powered pump supplying water for crop irrigation in India (Courtesy BP Archive)

deployment worldwide. In 1984, Solarex deployed a 21.6 kWp system in Egypt, delivering 41 m^3 per hour, which continued to operate for 30 years and 'turned the barren desert into a green paradise' [47]. Over the course of the 1980s, photovoltaic water pumping became an established technology, seeing use in many countries. In that decade, the European Union rolled out a Regional Solar Programme, bringing drinking water to the Sahel region; by 2009, 1 million people were benefitting from the supply [48]. In 2011, Grundfos was providing 10 000 photovoltaic water pumps per year. Figure 3.12 shows a typical solar-powered irrigation system in India.

In addition to supplying drinking water and irrigation, the ability to pump water enabled simple purification systems to be installed where water was abundant but undrinkable. In its most recent form, pumping has been used to drive a reverse-osmosis plant for water purification [49].

3.3.3 Consumer Electronics

The 1970s saw the rise of digital watches. Varadi recounts how, in 1974, he was contacted by a new-entrant watch manufacturer in the United States looking for a provider of photovoltaic cells as the new LED watches required more power than the conventional battery-powered mechanical ones [50]. A solar panel 20 × 20 mm consisting of eight silicon cells was required, which would have provided about 0.4 mW at 10% cell efficiency [50]. This was produced successfully, and 60 000 watches were sold. The same approach was quickly followed by a Swiss manufacturer. Power demand decreased as LCD watches took over from LEDs.

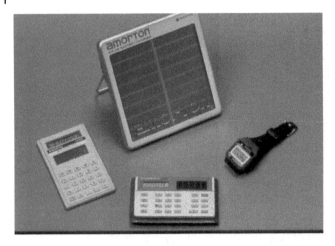

Figure 3.13 Typical solar-powered products from the early 1980s. *Source:* Panasonic Company

Alongside watches, handheld calculators also emerged, becoming a mass product in the 1970s. While early solar chargers were made with silicon cells, the Sanyo company, with its new amorphous silicon technology, found it cheaper and more aesthetically pleasing to use thin-film solar cells, and the use of crystalline silicon for such low-powered products declined. By the early 1980s, there was a range of products available, from watches, calculators, and radios to battery chargers and garden lights, all powered by small photovoltaic panels. Figure 3.13 shows a representative cross-section of consumer products from that time.

Although small in power, a large number of units were shipped, making this a truly commercially profitable market. A simple breakdown of the world market by product sector is given in Table 3.2. It is evident from this table that photovoltaic power for consumer products represented 30% of the total market and almost half of the truly commercial nongovernment sector. This is the reason for the 'Milliwatts' reference in the title of this book, as in this early period when the world market was only 25 MWp p.a., a very significant proportion of this was made up of low-powered devices of a few milliwatts each. It can also be seen from the table, however, that the market

Table 3.2 World market by sector in the mid-1980s

Application	1984		1985	
	MWp	% market	MWp	% market
Consumer product (<10 Wp)	6.65	27	7.2	29.6
Commercial	9.15	36	11.2	45.7
Government programmes	9.2	37	6.0	24.7
Total	24.0	100	24.4	100

Source: Based on P Maycock. PV News

grew very little from 1984 to 1985; this was exceptional, as through most of the 1980s growth was around 25% p.a. This was a consequence of the cancellation of tax credits for photovoltaics in the United States, which had been counted in the government figures.

3.4 Summary

This chapter has demonstrated how early markets arose for photovoltaic products. These were true market applications growing out of the real need for off-grid electrical power in difficult locations. It was this need, rather than technological advances and cost reductions by the manufacturers of photovoltaic modules, that caused market growth and validation of the photovoltaic technology. As chapter 5 will show, the on-grid market came to dominate global shipments and the off-grid market became insignificant in volume terms. Nevertheless, Wolfgang Palz has shown in his book *Solar Power for the World* how the off-grid domestic market has transformed lives, and how it continues to do so [51].

References

1 W. Callaghan et al.: Flat Plate Solar Array Project Final Report, October 1986, JPL Publication 86–31.
2 R.H. Blieden and F.H. Morse: Proceedings of the 10th IEEE PVSC (1973) 216–226.
3 W.R. Cherry: Proceedings of the 8th IEEE PVSC (1970) 331–337.
4 S. Susaki: IEEE Spectrum April 2014.
5 J.A. Duffie and G.O.G. Lof: Proceedings of the 6th IEEE PVSC (1967) 70–80.
6 J. Perlin: 'From Space to Earth' pub: Harvard University Press (2000) 61.
7 https://global.sharp/solar/en/history/#:~:text=In%201959%2C%20Sharp% 20became%20the,power%20plants%20around%20the%20world accessed 7 August 2020.
8 J. Perlin: 'From Space to Earth' pub: Harvard University Press (2000) 71–76.
9 P. Varadi: 'Sun Above the Horizon' pub: Pan Stanford (2014) 215–227.
10 https://www.trinityhouse.co.uk/about-us/a-to-z-of-trinity-house/solar-power accessed 7 August 2020.
11 J. Perlin: 'From Space to Earth' pub: Harvard University Press (2000) 85–101.
12 I. Garner: Proceedings of the 18th IEEE PVSC (1985) 275–279.
13 https://trove.nla.gov.au/work/22283525 accessed 7 August 2020.
14 Y. Chevalier et al.: Proceedings of the 15th IEEE PVSC (1981) 201–204.
15 P. Varadi: 'Sun Above the Horizon' pub: Pan Stanford (2014) 120.
16 J. Perlin: 'From Space to Earth' pub: Harvard University Press (2000) 99.
17 A.W. Penbody: 'Control of Pipeline Corrosion' 2nd edn pub: NACE International (2001) .
18 V. Ashworth: 'Corrosion' 3rd edn, vol. 2 pub: Elsevier (1994).

19 https://www.hubbell.com/gai-tronics/en/Products/Data-Communications/ Telephones/Rugged-Telephones/VoIP/NEMA-4X-VoIP-Rugged-Handset-Telephone-354-700-Series-Keypad-with-Armored-Cord/p/2965612 accessed 7 August 2020.

20 https://www.prolectric.co.uk/temporary-solar-lighting/protemp-solar-street-light/ accessed 7 August 2020.

21 J. Perlin: 'From Space to Earth' pub: Harvard University Press (2000) 64.

22 R.E. Tolbert and J.C. Arnett: Proceedings of the 18th IEEE PVSC (1985) 1149–1152.

23 H.J. Weger et al.: Proceedings of the 22nd IEEE PVSC (1991) 586–592.

24 J.H. Wohlgemuth: Proceedings of the 23rd IEEE PVSC (1993) 1090–1094.

25 http://www.clui.org/ludb/site/original-carrizo-solar-power-plant-site accessed 7 August 2020.

26 N.H. Holley et al.: 1st WCPEC (1994) 893–896.

27 R. Spencer et al.: Proceedings of the 17th IEEE PVSC (1984) 872–875.

28 G. Grassi: Proceedings of the 15th IEEE PVSC (1981) 871–875.

29 J. Perlin: 'From Space to Earth' pub: Harvard University Press (2000) 147.

30 J. Perlin: 'From Space to Earth' pub: Harvard University Press (2000) 149.

31 https://www.presidency.ucsb.edu/documents/the-environment-message-the-congress accessed 7 August 2020.

32 W. Palz: 'Solar Power for the World' pub: Pan Stanford (2014) 153.

33 W. Palz: 'Solar Power for the World' pub: Pan Stanford (2014) 67.

34 J. Perlin: 'From Space to Earth' pub: Harvard University Press (2000) 128.

35 W. Palz: 'Solar Power for the World' pub: Pan Stanford (2014) 157.

36 J. Perlin: 'Let It Shine : 6000 Year History of Solar Energy' pub: New World *Library* (2013) 403.

37 P. Varadi: 'Sun Above the Horizon' pub: Pan Stanford (2014) 121.

38 N. Williams in 'Solar Power for the World' ed. W. Palz pub: Pan Stanford (2014) 481–486.

39 A. Cabraal in 'Solar Power for the World' ed. W. Palz pub: Pan Stanford (2014) 453.

40 https://solar-aid.org/ accessed 7 August 2020.

41 United Nations World Water Development Report: Leaving No One Behind (2019) 10–33.

42 J. Perlin: 'From Space to Earth' pub: Harvard University Press (2000) 104.

43 W. Palz: 'Solar Power for the World' pub: Pan Stanford (2014) 158.

44 J. Perlin: 'From Space to Earth' pub: Harvard University Press (2000) 103.

45 P. Varadi: 'Sun Above the Horizon' pub: Pan Stanford (2014) 200.

46 J. Perlin: 'From Space to Earth' pub: Harvard University Press (2000) 115.

47 P. Varadi: 'Sun above the Horizon' pub: Pan Stanford (2014) 211.

48 P. Varadi: 'Sun above the Horizon' pub: Pan Stanford (2014) 203.

49 B. Roy in 'Solar Power for the World' ed. W. Palz pub: Pan Stanford (2014) 476.

50 P. Varadi: 'Sun Above the Horizon' pub: Pan Stanford (2014) 138.

51 W. Palz: 'Solar Power for the World' pub: Pan Stanford (2014) 109–174.

4

Silicon Technology Development to 2010

4.1 Introduction

Chapter 2 described the rise of the terrestrial solar industry. Silicon solar cells were the first commercial technology to be developed, and they have continued to be the dominant one to the present day. Competing technologies have not yet displaced them – as will be discussed at some length in Chapter 7. This chapter looks at the underlying reasons why silicon has been so robust and at the process changes that developed between 1980 and 2000 which enabled it to become a gigawatt-scale energy-generating technology.

4.2 Technologies Supplying the Global Market

Figure 4.1 shows how the product shipped to market year by year has varied with the cell technology deployed. In the early 1980s, the product was almost 100% monocrystalline silicon, with small quantities of ribbon silicon and multicrystalline silicon emerging. From 1982 onward, as described in Chapter 3, the growth of photovoltaics in consumer products gave rise to a significant market for the thin-film amorphous silicon technology. Through most of the decade, the consumer product market grew faster than any other applications, so that at its peak it achieved a 30% market share, albeit with a total shipment of around 10 MWp. The larger-scale power applications continued to grow while the consumer market stabilised so that thin-film share decreased and failed to develop significant applications in the larger power market, as will be discussed later in this chapter. By the early 2000s, the share of thin film in the now gigawatt-size market had receded to less than 10% [1]. Another surge occurred around 2010. This was in part due to the rapid growth in demand for crystalline silicon products, which outstripped the supply of semiconductor-grade polysilicon feedstock so that other cell technologies were deployed. In particular, First Solar started a large-volume supply of cadmium telluride modules at efficiencies around 10% and at a cost below that of silicon modules.

Photovoltaics from Milliwatts to Gigawatts: Understanding Market and Technology Drivers toward Terawatts, First Edition. Tim Bruton.

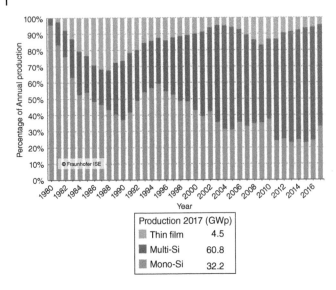

Figure 4.1 Global photovoltaics market by technology, 1980–2017(*Source:* Courtesy FhG.ISE)

Production 2017 (GWp)	
Thin film	4.5
Multi-Si	60.8
Mono-Si	32.2

The lower efficiency caused an increased balance-of-systems cost so that this was the most expensive technology to install, but it still offered a reasonable return given the feed-in-tariff (FIT) at the time (see Chapter 6) [2]. However, rapid investment in additional polysilicon capacity, initially in Germany and then later in China, de-bottlenecked the supply, and low-cost polysilicon fuelled a rapid growth of sales of silicon modules. Although First Solar continued to expand production, it did not keep pace with the global growth in demand and its market share fell. Start-up companies in copper indium gallium diselenide (CIGS) have yet to make an impact.

4.3 Advantages of Silicon as a Solar Cell Material

The dominance of silicon as a cell technology stemmed initially from the fact that it was almost the only semiconductor material used in the multibillion dollar market for electronic devices. The availability of semiconductor-grade silicon at Bell Labs had enabled the creation of the first solar cell, as described in Chapter 1. It has a number of other advantages as a solar cell material, as summarised in this section [3].

4.3.1 Readily Available

Silicon makes up 23% of the earth's crust, albeit largely in the form of silicates. Around 8 million tonnes of metallurgical-grade silicon, either as a pure metal or alloyed to iron to form ferrosilicon, are produced globally each year, at a price of around $3/kg [4]. Historically, a small proportion of the world supply was purified by the Siemens process to form semiconductor-grade silicon. At the start of the 1980s, this was around 20 000 tonnes p.a., but today it is in the region of 100 000 tonnes p.a., largely driven by photovoltaic demand, with China the leading manufacturer. The supply fully matches current demand, as reflected by the long-term decline in price, which fell below $8/kg in 2019 [5]. This contrasts with some other solar cell technologies, where the availability of elements such as indium and tellurium is seen as a limiting factor at the gigawatt scale of manufacture [6].

4.3.2 Elemental Semiconductor

Silicon is an element. This gives it advantages over other candidate materials for solar cells, as it is congruently melting at 1414 °C, and with a boiling point of 3265 °C [7] evaporation during crystal growth is not a significant issue. Fabrication of either monocrystalline or multicrystalline ingots is thus a simple process compared to III–V compounds, where generally the group V element is much more volatile, making crystal growth more complex. In a material like CIGS with four elements of widely varying boiling points and a wide range of possible compounds, control of stoichiometry over a large area is a major issue. This will be discussed further in Chapter 7.

4.3.3 Nontoxic

Silicon has a low toxicity and presents few hazards in either manufacture, field deployment, or end-of-life disposal [8]. This compares favourably with cadmium telluride, where both its constituent elements are known hazardous materials and require careful handling and disposal [9]. The nonhazardous nature of silicon also facilitates recycling from end-of-life modules [10].

4.3.4 Self-Passivating Oxide

It is well known that silicon immediately grows an oxide film on exposure to air. Given the right growth conditions and a clean silicon surface, a very low surface recombination of photocarriers can be achieved. This property is often used in the fabrication of high-efficiency silicon solar cells, although the refractive index of silica is too low for an optimum silicon solar cell antireflection coating (ARC). The highest-efficiency devices are therefore made with a very thin film of silica and a capping layer of another dielectric [11].

4.3.5 Synergy with the Global Semiconductor Industry

For the past 60 years, silicon has been the dominant semiconductor in the global electronics industry. Thus, from the earliest days – even from the initial work at Bell Telephone Laboratories – solar cell scientists have been able to make use of the massive existing infrastructure used to manufacture devices for the world's electronics and information technology industries (valued at $300 billion in 2019) [12]. Throughout the development of photovoltaic production, equipment for crystal growth, wafering, diffusion, and metallisation has been readily available – unlike in thin films, where production equipment has generally been built on a one-off basis. In addition, the extensive scientific knowledge base around semiconductor silicon has been hugely beneficial in the development of silicon solar cells. At the same time, as the photovoltaics industry has grown, it has diverged in some respects, and in some cases the mainstream electronics industry has benefitted from the developments in photovoltaic manufacturing.

4.4 Silicon Solar Cell Design Features

The theory of solar cells has been well described elsewhere [13,14], but in order to understand how the manufacturing technology developed between 1980 and 2000 it is necessary to describe some of the key principles of their operation. A schematic of a silicon solar cell is shown in Figure 4.2. The active part of the cell is a p/n junction in the silicon. The silicon wafer is usually p type, with resistivity typically $1\,\Omega/\text{cm}$ with boron doping (the base), and the surface is doped n type with phosphorus (the emitter). At the p/n junction in the non-illuminated state, electrons diffuse from the emitter to the base and holes diffuse from the base to the emitter until a strong enough electric field is built up to prevent further diffusion of carriers. The region where the electric field acts is known as the depletion region, as there are no free current carriers there. The electric field pushes electrons toward the n type region and holes toward the p type region. In operation, photons from the solar irradiation enter the silicon. If the photon energy is greater than the bandgap of silicon, an electron is excited from the valence band into the conduction band and a hole is created in the valence band. In silicon, the hole and the electron (the photocarriers) are free to move independently. If they are generated in the depletion region, the electron is pushed by the

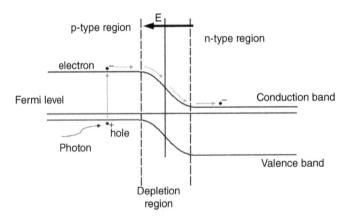

Figure 4.2   Schematic of a p/n junction solar cell, showing the flow of photocarriers (*Source:* Courtesy BP Archive)

electric field toward the emitter and the hole is ejected into the base. This rapid movement of charges from the depletion region creates a diffusion gradient of excess holes on the n side and electrons on the p side, causing the excess carriers to move to the depletion region and be separated across it. The emitter becomes negatively charged and the base positively charged, and if there is no external circuit, the process will continue until an electric field is generated, which prevents further diffusion of carriers across the depletion region. This field appears as a voltage across the terminals of the solar cell and is termed the 'open-circuit voltage' (Voc). If the terminals are short-circuited, there is no opposing electric field and all generated carriers flow across the depletion region; the current generated is normally exactly proportional to the number of incoming photons. This is termed the 'short-circuit current' (Isc).

The generation of the photocarriers is only one part of a cell's operation. The carriers must next be collected and delivered to the external circuit. This is done by providing low-resistance contacts to the front and back surfaces of the solar cell, as shown in Figure 4.3. The rear contact is simple and can extend over the full rear surface area. In its simplest version, this is a screen-printed silver/3% aluminium contact. The aluminium promotes good contact and is limited to 3% to allow solder contacts to the cell rear. The front contact is more complex. It is typically a grid, in order to allow as much light as possible to enter the cell. Fine grid lines extend over most of the surface area, but they connect to a wider contact strip for greater current carrying capacity, usually running the whole length of the cell. This is the bus bar. Early cells typically had two bars, but more recent larger-area ones require three or more to cope with the larger currents they experience (up to 8 A). Figure 4.3 shows an external load across the terminals of the solar cell. In either short- or open-circuit condition, the solar cell produces no real power. As the resistance of the load changes from zero at short circuit, the voltage across the terminals rises and power is dissipated in the load. As the load is increased, a point is reached where the maximum power is delivered it. This can be seen more clearly in Figure 4.4. The upper line shows the unilluminated diode

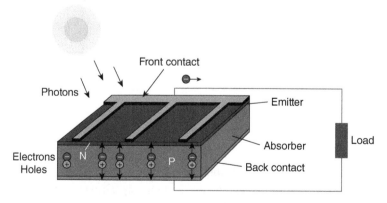

Figure 4.3 Essential features of a simple silicon solar cell (*Source:* Courtesy BP Archive)

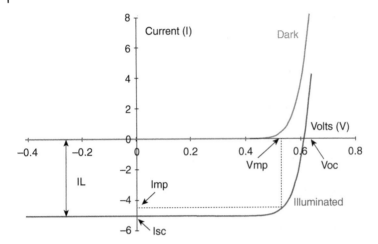

Figure 4.4 Dark (upper) and light (lower) current/voltage characteristics of a solar cell (*Source:* Courtesy BP Archive)

characteristic following the diode equation (where I and V are the corresponding current and voltage across the diode terminals, Io is the junction saturation current, e is the electronic charge, K is Boltzmann's constant, and T is temperature):

$$I = Io\left(e^{\frac{eV}{kT}} - 1 \right)$$

This follows a parabolic shape in the forward bias quadrant of the diode. In a good solar cell, it is assumed that the principle of superposition occurs and that a current of $-I_L$ (the current created by the incoming light) is applied to the dark current linearly across the whole voltage range. The maximum power point is achieved just as the current begins to drop rapidly toward Voc. The maximum power point is the product of Imp and Vmp. This gives rise to a definition of the cell's fill factor (FF) as:

$$FF = Vmp.Imp\Big/Voc.Isc$$

The fill factor is a useful parameter, as the maximum power of the cell is:

$$Pmax = FF.Isc.Voc$$

The solar cell efficiency (η) is then defined as:

$$\eta = Pmax\Big/A.G$$

where A is the cell area and G is the solar insolation per unit area on the cell.

It is thus obvious that in order to maximise the cell's efficiency, all parameters Isc, Voc, and FF must be increased. Isc is proportional to the number of photocarriers generated and collected at the cell terminals. It is lowered by reflection from the front

and rear surfaces of the solar cell, transmission of light through the cell without generation of carriers, and losses due to recombination of photocarriers [15]. Voc is funda-
mentally limited by the bandgap of the semiconductor, which is 1.1 V for silicon. This
value is reduced by the internal resistance of the solar cell and by recombination processes inside it [15]. In practice, Voc ranges between 600 mV for typical production cells
and 700 mV for the best laboratory cells. The reasons for these differences have been
discussed at length by Gray [16], who shows that the fundamental Auger recombination process limits the maximum Voc to between 720 and 750 mV, depending on the cell
thickness and minority carrier diffusion length (i.e. the distance an electron can travel
before recombining with a hole in p type material). Other recombination mechanisms,
particularly at the front and rear surfaces and at defects in the bulk of the wafer, further
reduce Voc, as does any shunting of the p/n junction. The fill factor is defined by the
Voc and the diode quality factor. However, it is reduced by increased series resistance in
the cell and by a low cell shunt resistance. The series resistance is made up largely of the
bulk resistance of the wafer and the resistance of the metal contacts. The shunt resistance arises from leakage paths through the p/n junction, usually caused by local defects
in the emitter through which the metallisation penetrates the base of the solar cell.
Series resistance should be low and shunt resistance high.

The route to high efficiency can be summarised as minimising sources of recombination, ensuring maximum generation of photocarriers, and minimising resistance
effects. The minority carrier diffusion length is a good indicator of the material quality
and the ultimate solar cell efficiency that can be achieved. It indicates the amount of
recombination within the bulk of the silicon wafer. It is limited by impurities and crystalline defects in the wafer.

4.5   Silicon Solar Cell Manufacturing from 1980 to 1990

The years from 1980 to 1990 saw a transformation in the manufacture of silicon solar
cells. As described in Chapter 3, at the start of the decade, a terrestrial solar cell
emerged out of space technology. This was generally a 100 mm round monocrystalline
cell based on a p type wafer of 500 μm thickness. The metallisation on the front side
was a screen-printed silver with an H pattern, with two bus bars and grid lines 300 μm
wide. The rear metallisation was a solid screen-printed metal back of silver/3% aluminium. The surface was textured, but normally did not have an ARC. A schematic of
a typical solar cell of the period is shown in Figure 4.5.

By 1990, the technology had changed. The wafers were now either fully square-
cast multicrystalline (mc-Si) wafers or pseudo-square monocrystalline (p-sq.-Si)
wafers, typically 100 m across the flats. The thickness was typically 300 μm and the
grid line width 200 μm. The mc-Si wafers had an ARC – usually APCVD-deposited
tantalum or titanium oxide, although some manufacturers used PECVD silicon
nitride [17]. The CZ p-sq.-Si cells in general did not. The rear contact was either a
silver grid or a complete coverage of the cell rear. Some manufacturers used a thick

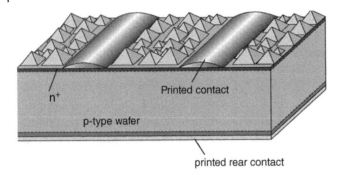

Figure 4.5 Typical solar cell from 1990 (*Source:* Courtesy BP Archive)

screen-printed aluminium rear to give a BSF [18] with silver pads allowing solder contacts. Module technology had not changed significantly.

This section describes how manufacturing techniques developed through this decade, setting the foundations for the rapid expansion that occurred in later years.

4.5.1 Silicon Feedstock

As silicon is an indirect semiconductor and therefore a poor absorber of light, silicon wafers have to be relatively thick to absorb all the incident solar irradiation. It has been calculated that 1.5 mm of silicon is necessary to absorb all the long-wavelength radiation above the silicon bandgap. Because the energy content of this irradiation is low, it is possible to use thinner wafers without a significant loss in efficiency. At the start of the 1980s, wafers were typically 500 μm thick, but this had reduced to 350 μm by 1990, and present technology uses wafers typically 180–200 μm in thickness. With 1980s technology, it was estimated that 19 tonnes of silicon per megawatt of solar cells was required [19]. At a typical price for that period of $40/kg of semiconductor-grade polysilicon feedstock (sg-Si), this equated to $0.76/Wp for feedstock alone. While the photovoltaic industry was developing through the 1980s and 1990s, there was no independent supply of high-purity silicon and the industry had to rely on reject material from the mainstream semiconductor industry. This took the form of off-specification polysilicon, tops and tails of Czochralski (CZ) boules, reject crystals, pot scrap, and monitor wafers. It was generally assumed that around 10% of the prime polysilicon would be available to the solar industry. This was typically available at around $20/kg, so the feedstock cost of photovoltaics was below $0.4/Wp. This scenario worked reasonably well up to around 1998, when the global production of semiconductor-grade polysilicon was around 20 000 tonnes p.a. [19], which was sufficient for around 120 MWp of annual photovoltaic production. However, the semiconductor industry was growing at around 3% p.a. while the photovoltaic industry was growing at around 25% p.a., and even higher growth rates would be achieved after 2000, necessitating dramatic changes in the feedstock supply.

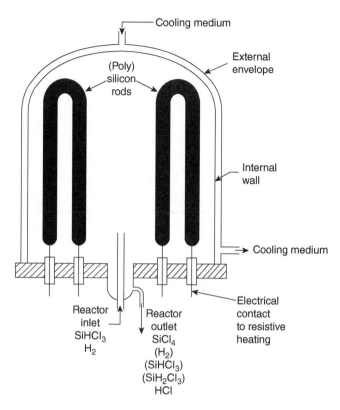

Figure 4.6 Schematic layout of a Siemens reactor used in the production of semiconductor-grade silicon [20] (*Source:* B. Ceccaroli and O. Lohne, "Handbook of Photovoltaics Science and Engineering" (2003) pub J Wiley, 179)

Up to 2000, the semiconductor industry was dominated by a small number of large companies, including Wacker Chemitronic (Germany), Hemlock Semiconductor (United States), and Tokuyama (Japan). These relied on the Siemens process for the production of polysilicon, which is described in some detail by Ceccaroli and Lohne [20], while a typical Siemens reactor is shown in Figure 4.6.

The starting point of the Siemens process is metallurgical-grade silicon, which is reacted with hydrochloric acid to form trichlorosilane:

$$Si + 3HCl = SiHCl_3 + H_2 \tag{4.1}$$

Trichlorosilane is a liquid at room temperature (boiling point 31.8 °C). It must be doubly fractionally distilled to achieve the necessary levels of purity. The highly purified trichlorosilane is then injected with high-purity hydrogen gas into a Siemens reactor, where silicon seed rods in a U shape are heated to a surface temperature of 1100 °C.

The following reaction (the reverse of 4.1) then occurs, reducing the trichlorosilane back to silicon:

$$2HSiCl_3 + H_2 = Si + 3HCl \tag{4.2}$$

However, competing reactions also occur, such as:

$$2\,HSiCl_3 = Si\,H_2Cl_2 + SiCl_4 \tag{4.3}$$

and:

$$HCl + HSiCl_3 = SiCl_4 + H_2 \tag{4.4}$$

This gives significant volumes of silicon tetrachloride as a waste product, which is often used in the production of fumed silica.

As can be seen from reactions 4.2, 4.3, and 4.4, the waste stream from the Siemens reactors contained a mixture of toxic and corrosive gases. The handling of this stream was an integral part of the process, and both Wacker and Hemlock were collocated with silicone plants which could utilise the waste products. Reaction 4.2 is highly endothermic, requiring large amounts of cheap electricity to be cost-effective. For this reason, polysilicon plants were often sited in places where cheap hydroelectricity was available. This also led to high amounts of embedded energy in the silicon solar cell production and to relatively long times being required before silicon photovoltaic systems could become net generators of electricity; recent studies have put the number at as low as 1.8 years in a good solar regime in Southern Europe [21].

Given the high cost, complexity, and handling difficulty of gaseous silicon compounds, alternatives to the Siemens process were sought early on. One of the first attempts was the development of a route from metallurgical-grade silicon to silane gas [22]. Dichlorosilane was formed catalytically from trichlorochlorosilane, as given in reaction 4.3 [20]. This was then catalytically reacted to produce silane as follows:

$$3\,Si\,H_2Cl_2 = 2\,HSiCl_3 + SiH_4 \tag{4.5}$$

The silane was then decomposed in a reactor similar to the Siemens reactor:

$$SiH_4 = 2H_2 + Si \tag{4.6}$$

This reaction is simpler than the Siemens process and gives a denser polysilicon rod with no toxic, corrosive gases at high temperature. A disadvantage is that the yield of silane in reaction 4.5 is relatively low and a constant recycling of the chlorosilane gases is essential. By 1986, the process in reaction 4.6 was scaled up to 1200 tonnes p.a. at a Union Carbide Corporation facility [22]. Following restructuring of Union Carbide, the ownership of the plant passed to Komatsu, and then finally to the Norwegian photovoltaic manufacturer REC. The process produced such high-quality polysilicon that it was mainly used in the semiconductor industry and did not achieve the cost break-through expected for photovoltaics.

It was quickly realised that metallurgical-grade silicon alone as a feedstock was not a prospective route, as the large number of silicon carbide particles shunted the solar cell [23] – although some cells were made with efficiencies as high 12.9% [24]. The low cost, ready availability, and much lower embedded energy combined to promote a pro-longed exploration of the possibility of using metallurgical-grade silicon directly as solar cell feedstock. The different approaches have recently been reviewed by Ceccaroli and Tronstad [25]. In the early 1980s, programmes were started in the United States, Japan, and Europe to try to utilise mg-Si [26] to produce an upgraded metallurgical-grade silicon suitable for solar use (UMG-Si). The main impurities were carbon, transition metals, aluminium, phosphorus, and boron, with the latter three being well-known dopants in silicon. A common theme of the upgrading process was gas treatment of the liquid silicon to reduce the dopant elements, followed by directional solidification mainly to remove silicon carbide particles and post-solidification crushing and acid leaching to remove metals. Various experiments were undertaken with high-purity quartz and carbon feedstocks [27]. Plasma treatment of molten silicon was developed in France [28], with an oxygen-containing argon plasma to oxidise the boron and phosphorus impurities. While there was a very broad approach to the topic, little com-mercial success was achieved, due to variability in the process and difficulties in scal-ing up the rather long gasification processes from the laboratory scale to the hundreds of tonnes required for commercial production. Dow Corning had begun research early in the 1970s [29], but the product was only commercialised in the 2000s, at the height of the silicon shortage, and was withdrawn in 2010. The major silicon producer Elkem also started work on UMG-Si in the 1970s, in collaboration with Exxon (owners of Solar Power Corporation) [25]. Elkem continued its work in conjunction with a num-ber of partners and eventually launched a commercial product around 2005. It was able to demonstrate that CZ wafers with 100% UMG-Si could achieve the same high efficiency of 18.7% as solar cells made from normal semiconductor-grade silicon [30]. Elkem currently has a production capacity of 6000 MT p.a. of its ESS UMG-Si [31].

Another early route to the use of UMG-Si involved fabricating wafers as low-cost substrates on which thick epitaxial layers of high-quality silicon could be grown. One development reported 12.8%-efficiency solar cells made with a 100 µm-thick epitaxial layer; this was comparable to the efficiencies obtained on heavily doped sg-Si sub-strates [32]. This gave rise to an enduring research programme looking at the use of low-cost silicon substrates with epitaxial silicon layers to produce high-efficiency solar cells. In 2009, it was reported that a 16.1%-efficiency epitaxial cell could be fabricated on an electrically inactive silicon substrate [33].

While UMG-Si has made progress, the rapid expansion of the photovoltaics industry from 2000 onward was largely accomplished with sg-Si formed by derivatives of the Siemens process. In a detailed MUSIC FM study of the route to sustainable high-volume 500 MWp p.a. photovoltaic production, it was concluded that a simplified

Siemens process could produce high-quality low-cost polysilicon feedstock for €20/kg [34]. It was apparent by 2006 that a major shortage of polysilicon was occurring [35]. 2006 was also the year in which polysilicon usage in the photovoltaics industry first matched that of the global semiconductor industry. If demand for the growing photovoltaics industry was to be met, therefore, the global supply had to grow from 35 000 tonnes p.a. in 2006 to at least 100 000 tonnes p.a. by 2010 [35]. The established industry suppliers promised new technology to meet this demand, but it was slow in coming [36–38], and there was a significant shortfall, with spot prices for polysilicon rising to over $300/kg. Historically, it was assumed that the existing suppliers could meet demand and it was generally believed that a polysilicon production plant had to be integrated with manufacturing facilities for other silicon products, such as silicones. However, GT Solar (now GT Advanced Technologies, GTAT) began to supply turnkey polysilicon plants, enabling start-up companies – particularly in China – to begin polysilicon production. Most recently, GTAT has received an order for a 12 000 tonne p.a. polysilicon plant in Asia. This plant will use FBR technology, which has lower throughput costs [39]. China now dominates global production of polysilicon, with a current market spot price of $8.8/kg [40].

4.5.2 Crystallisation

In 1980, the dominant technology used in the crystallisation of silicon was the CZ approach (see Figure 4.1). As described in Chapter 3, this came about because this was the only available commercially available source of adequate-quality silicon wafers. The CZ approach had high-cost elements, however. The polysilicon feedstock charge could not be fully converted into useful single-crystal material. Conical sections at the beginning and end of the growth process could not be used, although they could be recycled. Something like 10% of the feedstock charge was necessarily left in the crucible (pot scrap). Some pot scrap could be recycled, but it contained all the impurities rejected from the growing crystal. The growth rate was low, typically at 3 cm/s.

It was realised early on that multicrystalline silicon offered a potentially lower-cost route. As early as 1955, Chapin attempted to make solar cells with cast multicrystalline silicon [41], but his experiments were unsuccessful due to the small grain sizes used. The demand for lower-cost wafers in the 1970s reignited interest in this material. Three independent approaches were followed. In Europe, the Wacker Company had already developed large-grained cast multicrystalline silicon for other applications [42], and AEG GmbH – which had been making silicon space solar cells since 1968 – successfully used this material in the production of solar cells. These cells were typically 10% efficient on 100 cm^2 wafers and 14.5% on 4 cm^2 [43]. In the United States, Lindmayer at Solarex was experimenting with cast silicon [44]. He demonstrated that, provided the grain size was greater than a few millimetres, 10%-efficient solar cells could be made. Since the wafers were typically 0.5 mm thick, this meant that there should be no grain boundaries across the current flow in the cell. A grain boundary normal to the current flow would greatly increase recombination in the cell and become a source of high resistance, as observed by Chapin. Both Wacker and Solarex were melting silicon in a crucible at this time, before pouring it into a mould to form an ingot (see Figure 4.7). The disadvantage of this process was that heat flow was three dimensional, so that while large grains could be produced, columnar grains could not be generated through the whole ingot. Figure 4.8 shows a typical cross-section of a Wacker Silso ingot. The Silso material consisted of a core of relatively high-lifetime material surrounded by a ring of fine-grained low-lifetime material.

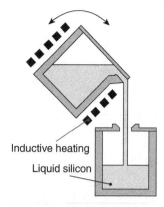

Inductive heating

Liquid silicon

Figure 4.7 Wacker Silso casting furnace (Courtesy Wiley)

In the third approach, in parallel with the work at Solarex and Wacker, significant advances were made in the directional solidification of silicon at Crystal Systems (CSI) [45]. This company had been founded in 1971 to exploit the heat exchange method (HEM) for the production of single-crystal aluminium oxide. In this method, heat was extracted through a heat exchanger at the base of a crucible so that growth could be initiated from a seed crystal, which was kept from melting by a cooling mechanism. In 1976, CSI received funding from JPL to develop this technology for silicon. The project was initially directed toward producing single-crystal silicon. This was successful, but it was found that the multicrystalline material – with good columnar crystal growth and grains up to 1 cm in size – produced cells with similar efficiency to the monocrystalline product

Figure 4.8 Cross-section of part of a Wacker Silso ingot (Courtesy T. Bruton)

Figure 4.9 Cross-section of a directionally solidified multicrystalline silicon ingot (Courtesy BP Archive)

via a simpler process [46]. A typical cross-section from a cropped bar is shown in Figure 4.9, where grains of 1 cm dimension grow over half the height of the brick. Growth was rapid at the base following nucleation, giving a very fine-grained material. It then stabilised, and large grains emerged. At the top of the ingot, the majority of the impurities were incorporated into the silicon, making growth unstable due to constitutional supercooling. Fine-grained material again developed. As shown in Figure 4.10, this led to low-lifetime regions at the top and bottom of the ingot, with a plateau of high-lifetime material across its major part. The associated solar cell efficiency is also

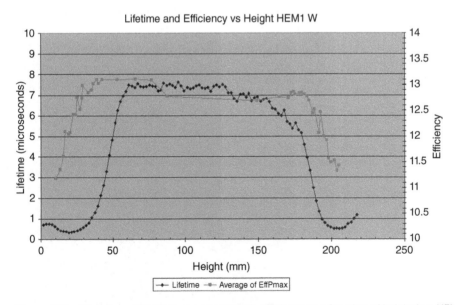

Figure 4.10 Minority carrier lifetime and solar cell efficiency as a function of height in a HEM directionally solidified silicon ingot (*Source:* Courtesy BP Archive)

shown in Figure 4.10. It can be seen that the zone of good efficiency is wider than that of good lifetime. This is due to beneficial effects in the cell processing, including phosphorus and aluminium gettering. This will be discussed in more detail later. Initially, the peak lifetime was just over 7 µs, but process optimisation by the end of the 1990s meant lifetimes as high as 18 µs could be achieved [47].

Directional solidification occurs in square fused silica crucibles. At the melting point of silicon, a strong interaction between the molten substance and the silica crucible takes place, leading to strong adhesion between the ingot and the crucible on solidification. The coefficient of expansion of silicon is much higher than that of fused silica, so that strong forces can be exerted on the ingot by the crucible and explosive shattering of the ingot can occur [45]. CSI reduces this effect by using a vacuum atmosphere for melting, which reduces the interaction, and pre-glazing the crucible to cause a weakened surface layer, which devitrifies the crucible, causing it rather than the ingot to break. While this approach worked reasonably well with ingots up to 30 kg in weight, it proved difficult to scale up beyond this, as discovered by BP Solar in its Andover, Massachusetts plant in the early 1990s [48]. The UK company Crystalox began making directional solidification equipment in the late 1980s, but it used an argon atmosphere and a silicon nitride-coated crucible. The silicon nitride prevented adhesion of the ingot to the crucible, so that large ingots could be grown (a 60 kg ingot with dimensions 45×45 cm was typical).

Directional solidification then became the dominant technology. The Wacker Silso process had been scaled up at its facility in Freiberg, Germany, which passed into Bayer ownership in 1994, and then into the ownership of Solar World in 2000. By 2002, HEM furnaces were ordered to replace the Silso technology [49]. Thus the directional solidification process had become the dominant route to multicrystalline silicon.

Having shipped over 3700 systems, GT Solar currently offers a DSS850 system casting ingots up to 820 kg in weight with dimensions 100×100 cm and a height of 35 cm [50]. A typical value for the lifetime of wafers taken from a 800 kg ingot is 7 µs [51].

While the multicrystalline silicon approach is generally lower-cost than monocrystalline CZ silicon, its solar cell efficiency is also lower. There has therefore been a push to increase the efficiency of solar cells using mc-Si. This has come to be known as HPM (high-performing material) mc-Si, and is based on control of nucleation of the silicon followed by promotion of favourable directions of grain growth in order to give low recombination grain boundaries and a minimum number of dislocations within the grain. Work done at BP Solar revisited the original CSI work [46] in order to grow near-single-crystal ingots through directional solidification [52]. It was demonstrated that solar cells made from cast near-monocrystalline silicon could produce cell efficiencies of 18.0% – close to that of standard CZ silicon, at 18.4% [53]. At the same time, it was shown that controlling defect generation from initial dendrite nucleation could produce a high-quality material [54]. Further development in controlling grain growth in 600 kg ingot casting showed that 17.8%-efficient mc-Si solar cells (averaged over the ingot) could be produced on industrial production lines – again, close to the efficiency achievable with CZ silicon cells [55]. More recent studies have shown that 19%

efficiency in HPM silicon is possible, and that an n type material with a minority carrier lifetime of up to 330 μs can be produced [56].

Despite the economic advantages of mc-Si, CZ has maintained a significant share fraction of production, as its ability to produce higher-efficiency solar cells is important in the current market, where balance-of-systems costs are as significant as module costs and therefore higher-efficiency cells are cost-effective at the system level.

4.5.3 Wafering

At the start of the 1980s, the only production method for wafering silicon was the internal-diameter diamond saw used as standard in the semiconductor industry. One wafer was cut at a time using a tensioned diamond-coated steel blade. The mechanical requirements were such that the diamond blade was typically 300 μm thick and the kerf loss was around 350 μm. It was also a slow process, taking typically three minutes per wafer. While this continued to be the preferred technique through much of the 1980s, it was recognised that it led to a high loss of valuable crystallised silicon, so alternatives were sought. Schmidt [45] records a diamond-coated wire approach was adopted early on at CSI, with a reciprocating frame holding a number of parallel wires. This did not achieve commercial success. A more promising approach followed work initiated at the Solarex subsidiary, Intersemix, in Switzerland [57]; this led to the formation of a new company, HCT, which developed the first successful wire saw for cutting multicrystalline silicon bars. The innovation involved making use of long lengths of steel wire (up to 200 km) developed for the car tyre industry, which removed the need for a reciprocating table and allowed continuous cutting in one direction, with the wire running at a higher speed than was possible in a reciprocation machine. The first machine produced on this model is shown in Figure 4.11. The cutting medium was a slurry of silicon carbide particles of 10–20 μm dimension in a mineral oil carrier. The steel wire was initially 180 μm in diameter, giving rise to a kerf loss around 200 μm. As illustrated in Figure 4.12, several hundred wafers could be cut simultaneously. As the machine was optimised, a single multiwire saw was produced with the capacity of 24 diamond internal-diameter saws, with lower kerf loss and less surface damage. The mineral oil used as the carrier was soon replaced with ethylene glycol for easier cleaning and recycling of the cutting medium. The mechanism of the cutting has been investigated at length [58]. The first saw was shipped in 1987, and 100 were shipped by 1997. This became the standard technology in the photovoltaics industry. Meyer Berger also started shipment of multiwire saws.

In more recent times, an alternative to the silicon carbide slurry approach has emerged. While cutting with fixed diamond wires was first developed in the 1970s [45] but not pursued, the high cost of silicon led to this approach being revisited. By modifying the multiwire saw with long diamond-coated wires and reciprocating the direction of travel, good cutting has now been achieved [59]. The diamond wire saw is still under development, and is chiefly used for CZ silicon, while the slurry approach continues to be used for mc-Si [60]. This is because CZ silicon can be texture-etched

Figure 4.11   First HCT wire saw
(Courtesy Applied Materials Corp.)

Figure 4.12   Conceptual image of a
silicon car being cut by multiple wires
(*Source:* Courtesy Applied Materials
Corp)

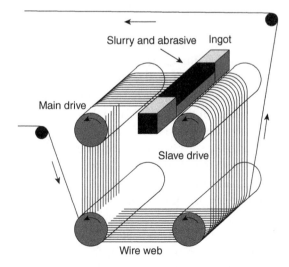

anisotropically to produce a textured surface, as described later, while mc-Si relies on the surface damage induced by slurry cutting to nucleate texture etching. A diamond wire does not produce the same degree of surface damage. Currently, 100 μm wires are used in slurry sawing, while wires down to 60 μm can be used for diamond sawing. The diamond wire approach has the benefit of providing faster cutting, with the potential to recycle the silicon kerf [61]. Recycling of the kerf has been a goal for many years. The finely divided nature of the silicon particles and contamination from silicon carbide particles, iron, and other transition metals have made this a major challenge, with no commercial process viable at this time.

4.5.4 Antireflection Coating

For good solar cell performance, the maximum amount of incident solar radiation should be coupled into the cell. A planar silicon surface will reflect about 30% of incident light. Surface texturing using the anisotropic etching of silicon works well for monocrystalline silicon but not for multicrystalline silicon [62]. Textured silicon will still reflect about 10% of incident light. The application of an ARC further reduces this in CZ-Si and was essential for mc-Si until acidic isotexturing of multicrystalline silicon was developed in the early 1990s. For an ARC on silicon, a quarter-wavelength film thickness of 600 nm was normally used. The refractive index of the ARC should be 2.4 for an unencapsulated cell or 2.0 for encapsulation under glass. The most commonly used process involved coating the wafers with either titanium dioxide or tantalum pentoxide using a metal organic source in an APCVD reactor. It was demonstrated that the thick-film contact could be fired through a titanium dioxide ARC as early as 1976 [63]. While these materials provided an adequate ARC, they did not offer any surface passivation. Silicon dioxide is passivating, but its refractive index is too low for an effective ARC. The significant breakthrough in the development of the ARC was the demonstration in 1984 of the use of a PECVD deposition of silicon nitride on to a multicrystalline silicon solar cell [64]. The silicon nitride coating fulfilled a number of functions. It acted as an effective ARC, with a refractive index between 1.95 and 2.05 depending on the deposition conditions. It acted as a surface passivation layer for the n type emitter, as H+ charges were trapped in the nitride layer and formed a positive field, repelling holes from the surface. It also acted as a source of hydrogen passivation grain boundaries well into the bulk of the wafer [65]. In the original work, photolithography was used to open windows through the nitride so that the silver screen-printed metallisation could make contact to the solar cell's emitter. Later work by BP Solar showed that it was possible to fire through the silicon nitride layer, greatly simplifying the process [66].

The challenge from an industrial perspective was to carry out the PECVD process in high-throughput equipment. The only reliable production equipment through the 1990s was a parallel-plate reactor, which could hold 60 100×100 m wafers in a 45-minute cycle.

A further breakthrough came when it was recognised that using a remote PECVD plasma protected the silicon surface from plasma damage and greatly increased the effective surface passivation, with surface recombination velocities as low as 4 cm/s

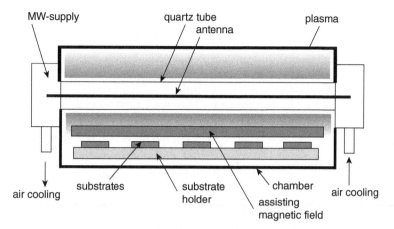

Figure 4.13   Schematic inline remote-plasma PECVD system for silicon nitride deposition on silicon wafers [69] (*Source:* W.J. Soppe et al., Prog in Photovolt. Res . and Appl. 13, (2005) 551–569)

reported [67]. Production equipment was required for the new approach, and an in-line system was manufactured by the Roth and Rau company, utilising technology developed at ECN [68]. Other manufacturers such as Centrotherm and OTB marketed similar systems. A schematic of the system is shown in Figure 4.13.

The ready availability and excellent performance of remote plasma led to this being the technology of choice for ARC deposition, as described in the next section.

4.6   Solar Cell Development to 2000

At the start of the 1980s, global production was just a few megawatts, with a large number of small companies around the world producing expensive solar cells in small volumes. By 2000, the global market was still small at 150 MWp, but the major technologies had been developed which would enable dramatic growth in the following decade. For most of this period, the development of CZ silicon cells and mc-Si followed different development routes, and they will thus be described separately. The development of the advanced solar cell structures is described in Chapter 8.

4.6.1   CZ Cell Development

A good description of the basic solar cell process in place in the early 1980s is given by Godfrey et al. [70] for the Tideland Energy plant in Sydney, Australia. Tideland was acquired by BP Solar in 1985. The plant utilised 75 mm round monocrystalline wafers of 500 µm thickness. The process sequence ran as follows:

1) *Saw damage removal.* This was normally an etch in 30% sodium hydroxide solution at 80 °C, removing around 20–30 µm of silicon for diamond internal-diameter-sawn wafers but less than 20 µm for wire-sawn wafers, where the saw damage was lower.

2) *Texture etching.* This was carried out at 90 °C in a 2% solution of sodium hydroxide with 2% isopropyl alcohol. Less than 10 µm was etched away in forming the (111) pyramidal surface texture.

3) *Surface cleaning and drying.* The wafers were etched in a 10% hydrofluoric acid solution to move any residual sodium silicate or silica and then a hydrochloric acid bath to remove any residual metallic impurities. They were then spun dry.

4) *n type diffusion.* This was carried out by spinning on a phosphorus source and driving the phosphorus in at between 900 and 1000 °C in nitrogen in a belt furnace to achieve a sheet resistance of around 30 Ω per square. Some other companies used a spray-on source of phosphoric acid, while others retained the traditional tube furnace diffusion.

5) *Phosphorus glass removal.* After diffusion, the wafer was coated with a phosphorous silicate glass (PSG). This was then removed with a few seconds' dip in 10% hydrofluoric acid, followed by washing and drying.

6) *Rear contact formation.* A thick film of aluminium was screen printed on the rear surface and fired up to 800 °C. Other companies either printed a silver contact with 3% aluminium to promote Ohmic contact to the p type silicon wafer or carried out a double printing firing step, where first silver pads were deposited and then an aluminium pattern with windows was applied to leave the silver exposed for subsequent soldering.

7) *Front contact firing.* A silver contact was screen printed on the front diffused side of the wafer in an H pattern, typically with 300 µm-wide grid lines at 3 mm spacing as the horizontal lines and two 2.5 mm-wide bus bars as the vertical ones. An LBIC scan of one such cell is shown in Figure 4.14. The metallisation where no current is generated is shown clearly.

8) *Edge isolation.* The phosphorus diffusion spread from the front of the wafer, and some diffusion occurred on the rear surface such that leakage ran from the front to the rear contact when the cell was fully contacted through this so-called 'parasitic junction'. In Tideland's case, a laser was used to scribe a trench around the periphery of the front surface, cutting through the junction and isolating it. At other

Figure 4.14 LBIC scan of a 1980s-type 100 mm round monocrystalline cell (Courtesy BP Archive)

companies, the wafers were coin stacked after diffusion and plasma etched in a carbon tetrafluoride/oxygen mixture, which etched the edge of the wafer to 1 µm or so depth, removing the parasitic junction.

9) *Cell test.* The cells were measured in typically home-built equipment using filtered quartz halogen lamps. Usually, this was done with separate current and voltage probes to the front surface bus bar, but contact was made to the whole rear surface so that any series resistance issues in the rear contact were masked. For accurate measurements, it was essential to have a good calibrated reference cell, well matched to the cells being tested, or else serious errors could arise as the spectrum from the lamps was very different to the AM1.5 Global spectrum recognised for standard solar cell measurements.

This cell process could reproducibly fabricate cells of between 12 and 14% efficiency. However, as discussed by Godfrey et al. [70], it had a number of drawbacks which limited the diffusion solar cell efficiency. Diffusion was not adequately controlled in the innovative infrared belt furnace used, and although 30 Ω per square was targeted for diffusion, it was often driven in harder to 15 Ω per square in order to achieve a consistent product [71]. It was well known that heavy diffusion limited the short-circuit current, as phosphorus silicide particles formed when strongly diffused junctions were cooled from diffusion to room temperature [72]. To avoid these effects, some work was done on ion implantation, and a very high 18% solar cell efficiency was achieved on a phosphorus-implanted cell in 1984 [73]. Subsequent research focussed on achieving lighter and lighter sheet diffusion. This put a constraint on the firing of the front contact. In order to produce an Ohmic contact, a silver silicon eutectic was formed, which penetrated into the emitter. The lighter the emitter diffusion, the thinner the emitter. Penetration of the eutectic through the junction caused cell shunting and loss of efficiency. By 1997, with developments on silver paste and control of the diffusion process, it was possible to use emitters as lightly doped as 80 Ω per square [74]. An alternative approach involves the use of selective emitters; this will be discussed in Chapter 8.

The formation of the back surface field (BSF) was also challenging. Although efficiencies as high as 14% could be achieved, the process was not consistent [71]. In addition, the work to make the rear surface suitable for soldering was time-consuming and caused wafer breakage. Tideland moved to an all-silver back, which was expensive, and then to a gridded-silver one, to reduce costs. A typical rear silver grid is shown in Figure 4.15. Most other companies also followed this approach, except for Sharp Corp., which consistently delivered 15%-efficient solar cell with an aluminium BSF [75].

In addition to the problems caused by the penetration of the silicon silver eutectic, the actual printing of the front contact was a challenge to the industry. In the 1980s, the print was limited by the quality of the screens available and the properties of the silver printing inks. Grid line widths of 300 µm were typical. The early cell in Figure 4.14 lost 15% of the active surface due to metallisation. Using lighter diffusions increased the series resistance in the emitter, so that the grid lines needed to be closer together. This required the grid lines to be narrower, in order to avoid increased shading. With

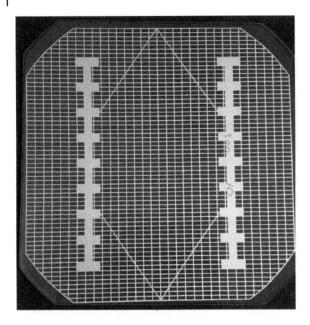

Figure 4.15 Rear silver grid pattern on a pseudo square CZ wafer (Courtesy BP Archive)

time, progress occurred, so that by 1997 it was possible to consistently print grid lines of 100 µm thickness [76]. While narrower lines may be printed, problems can occur in high-throughput production lines, with blocked screens and breaks giving high series resistance. The grid line has to carry a significant current, and the paste rheology limits the height of the fired paste to around 10 µm. Reducing the grid line width reduces the current carrying capacity, and this is another limitation; double printing of the front contact has been investigated, but this comes at the expense of reduced throughput [77].

4.6.2 Multicrystalline Silicon Processing

The multicrystalline silicon solar cell process broadly followed the CZ wafer one. An early process description at Photowatt International SA (France) is given by Donon [78]. The major differences from CZ processing were that no aluminium BSF treatment was used and a titanium oxide ARC was sprayed on instead of texture etching. Solar cell efficiencies around 10% were reported. While similar advances were made in diffusion and grid lines, the great challenge was to improve the antireflection properties, as the anisotropic etching used for monocrystalline material relied on the different etching rates of (100) and (111) planes. In the multicrystalline material, many other orientations were present, and only partial texture etching occurred. One early alternative was to use deep V grooves in the surface made by mechanical scribing [79]. While photolithographic techniques could be used in the laboratory and were successful in producing large-area cells of 225 cm^2 with efficiencies of 16.0%, the approach was prohibitively expensive [80]. Reactive ion etching using a chlorine plasma was pursued, and

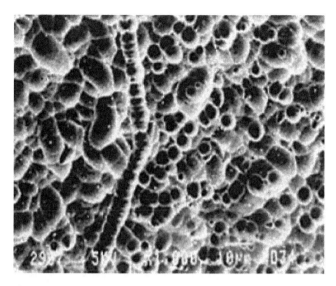

Figure 4.16 Textured surface of an acid-etched multicrystalline wafer [82] (K. Shirasawa, 2002. Mass production technology for multicrystalline Si solar cells. Prog in Photovolt. Res . and Appl. 10, 107–118. © John Wiley & Sons, Inc)

successfully created a textured surface. It was capable of producing solar cells of 17% efficiency on 225 cm^2 mc-Si wafers in the laboratory, but it never became an industrial process. The more successful route was acidic etching using hydrofluoric and nitric acid solutions [81]. Although this was an isotropic etch, it preferentially attacked the high stress points caused by wire sawing, inducing a surface texture as shown in Figure 4.16.

The simplicity and ease of introduction of this process soon led to its becoming the dominant production technology for mc-Si. By 2005, equipment companies such as RENA were manufacturing in line high-volume tools for the process [83].

4.6.3 Integration of Mono- and Multicrystalline Silicon Processes

As already described, mono- and mc-Si followed different production routes, and normally each cell manufacturer concentrated on manufacturing one wafer type – although for a period between 1987 and 1994, BP Solar manufactured both types in order to make use of all the commercially available wafers. A number of processes were used by different manufacturers: some used an aluminium BSF, some a nonpassivating ARC, and some a PECVD silicon nitride. At the end of the 1990s, the processes were integrated to become a standard manufacturing process. The defining work was carried out at IMEC [84]. As already described, PECVD silicon nitride is an ideal ARC which also passivates the surface. Firing through the silicon nitride with a screen-printed contact had been previously demonstrated and required a higher temperature and longer time than contacting directly to silicon. This was beneficial

Table 4.1 Improved process sequence, co-firing Al BSF and SiNx fire-through

Old process sequence	New process sequence
1. Saw damage removal	1. Saw damage removal
2. Texturisation	2. Texturisation
3. Deep diffusion	3. Diffusion
4. Selective etch back	4. PECVD SiNx
5. Oxide passivation	5. Screen-printed contacts
6. Al-BSF alloying	6. Parasitic junction removal
7. Al gettering	
8. Masking paste printing	
9. PECVD SiNx	
10. PECVD SiNx liftoff	
11. Screen-printed contacts	
12. Parasitic junction removal	

Source: J. Szlufcik et al., Proc 14th EUPVSEC (1997) 380–383 [84]

when using an aluminium paste to form a BSF, which required a higher temperature than was optimal for firing the front silver contact, necessitating the use of two firing sequences. The IMEC work optimised the firing conditions such that when firing through a silicon nitride ARC, a BSF could simultaneously be formed, reducing the number of process steps from twelve to six, as shown in Table 4.1. The new sequence was not only simpler but gave improved solar cell performance. For $100\,cm^2$ mc-Si wafers, the efficiency improved from 10.3% with APCVD TiO_2 to 13.7% with SiNx and 16.4% with additional V groove ARC. With CZ-Si on $100\,cm^2$, the efficiency gain was from 15.3 to 17.3%, replacing the TiO_2 with SiNx.

In this way, a process was developed which could be used equally well with both mono- and multicrystalline silicon. The only change was in the composition of the texture-etching bath. This led to process equipment manufacturers such as GTAM, Spire Corporation, and Centrotherm being able to offer turnkey production lines. For example, Centrotherm offered complete cell lines from 2000 [85]. Many start-ups were also able to begin production, particularly in East Asia, without the need for the investment in R&D that companies in Europe and the United States had made. It should be noted that even after 20 years of development, the monocrystalline silicon solar cell still had significantly higher efficiency compared to mc-Si.

4.6.4 Other Process Technology Changes

While the main changes relating to cell technology have already been described, some other notable changes also occurred. In 1980, cells were generally 75 mm and round. By the mid-1980s, they were 100 mm and square. The CZ wafer had rounded edges, being cut from a 150 mm boule, and thus was termed a 'pseudo square' (see Figure 4.15). Mc-Si was fully square. By the early 1990s, 125 mm had become the standard wafer size [34], and by 2000, 150 cm wafers were the norm. Wafer thickness also decreased with time. At the start of the 1980s, wafers were 500 μm thick, as this was the standard semiconductor thickness at that time, but this quickly decreased to 350 μm. It was assumed that the processing of 100 μm-thick wafers (or even thinner) would become industrially feasible. Demonstrations were made to show that a 20.4%-efficient solar cell could be made on 72 μm-thick wafers (4 cm^2) [86]. However, it proved difficult to process very thin wafers, particularly at large sizes, and 180–200 μm remains the standard for industrial production.

One process change that the introduction of thinner wafers stimulated was a change in the equipment used for wafer drying. In the 1980s, spin-rinser driers were the norm, where wafers in cassettes were spun at several hundred revolutions per minute and blown dry with hot nitrogen. This gave the wafers a large mechanical shock and was frequently a cause of wafer breakage. Thinner wafers were instead given a hydrofluoric acid dip in order to produce a hydrophilic surface, rinsed in deionised water, and drawn through an air blade, which removed all the water, yielding clean, dry wafers [87].

Improvements also occurred in screen printing, with the introduction of highly automated high-throughput machines that put the minimum of stress on each wafer. Early screen printers had used mechanical stops to position the wafer below the screen. The Baccini printer instead moved the screen to the wafer, minimising edge contact and reducing breakage. Modern versions can process 1250 wafers per hour and have a breakage rate of less than 0.2% [88].

In the 1980s, wafer-to-cell yields of 90% were common, but over time this increased to better than 97% [34].

4.7 Module Technology

Most of the innovation in module technology came in the early 1980s, as described in Chapter 2. Cells needed to be connected in series to generate a useful voltage. For a single silicon solar cell, the maximum power point occurred just below 0.5 V. In the early applications of off-grid systems, a voltage above 17 V under standard test conditions (STC) was required in order to successfully charge lead acid battery packs in high-temperature regions. This necessitated 36 cells connected in series, and this was thus the standard module size until a significant on-grid market emerged in the 1990s. The preferred manufacturing technique was to solder a strip approximately twice the length of the cell to each bus bar. This was usually a semi-hard copper strip, 3 mm wide in the early days, coated with a lead/tin solar alloy. It was soldered to the front surface of the cell, usually by hand, either completely along the length or at a number of points. This was called the tabbing process, and the strip was known as the tab. To form the module, the 36 cells were laid face down and the long lengths of tab were

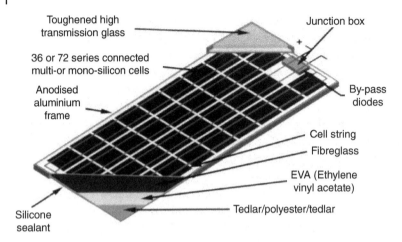

Figure 4.17 Typical module construction (*Source:* Courtesy BP Archive)

brought from the front of one cell to the rear of the next and soldered to form the series interconnection. The preferred layout was four strings of nine cells, with additional tabbing strips joining the strings together. This was termed stringing. Often, a stress-relieving kink was put in to the tabbing between cells to allow it to flex as the solar cell expanded and contracted in the daily cycle of use. To form the finished module, a lay-up procedure was used. The structure is shown in Figure 4.17. A textured, toughened, low-iron glass superstrate was used as the front side. This was placed with the smooth glass side downward and then covered with a sheet of the EVA encapsulant followed by a sheet of fibreglass to allow air extraction in the vacuum lamination step. The cell strings were laid on to the EVA and a further sheet of EVA was placed over them. This sheet had holes to allow the tab connections from the cell strings to be brought out to the junction box. Finally, the back sheet was applied. This was a laminate of Tedlar/polyester/Tedlar. The Tedlar gave UV stability, while the polyester acted as a moisture barrier. In the early days, an aluminium foil was used instead of polyester as it had better moisture resistance, but it was a source of electrical failure when solder bumps penetrated the back sheet from the soldered string and so was replaced by polyester.

The actual lamination was done in a vacuum chamber with a heated platen and a flexible diaphragm. The laminate was placed glass side down on the heated platen and then heated with air in the chamber to 80 °C, at which point the EVA softened. The vacuum was then applied to the chamber, and the diaphragm – with atmospheric pressure on one side and vacuum on the other (the laminated side) – was pressed down on the laminate, which was heated to 150 °C. The EVA melted and crosslinked while the air was extracted by the vacuum. Once crosslinking was achieved, resolidifying the EVA, the laminate was cooled and air was admitted back in to the chamber. The laminate was then trimmed and a junction box was fitted to its rear, with contacts made through the back sheet and EVA to the cell string, arranged so that bypass diodes could be fitted across two 18-cell strings. The bypass diodes were configured so that if a string contained a broken cell or was significantly shaded, rather than going into reverse bias and acting as a load to other modules in the series, the diodes began conducting, limiting power loss in the module string overall. Without them, large amounts of power could be dissipated in the defective string, causing irreversible damage such as melting of the EVA, bubbling, or major delamination.

This design has remained substantially unchanged to the present day. Improvements have been made to the UV stability of the EVA, the lead tin solder has largely been replaced by a silver/lead/tin eutectic in order to reduce the lead content of the module [89], and alternative encapsulants such as polyolefins (which are thermosetting rather than crosslinking, and are therefore much more easily recyclable) are under development, which show some better resistance to degradation mechanisms compared to EVA [90]. The only other major change has been in the automation of the process. While early operations were hand-soldered, a full range of automated equipment is now available, such that module assembly is virtually a non-manual operation today [91].

While glass/EVA/TPT remains the standard, glass/glass modules are frequently used in BIPV applications. Although EVA could be used, the preferred method has been to use a liquid resin infill to encapsulate the cell string. In the early years, silicone or polyurethane was used, but over time these were replaced by polyacrylate, as favoured by the building industry for use in double-glazed units [92].

4.8 Summary

This chapter has described how the technology for silicon solar cell production matured to became the cornerstone of the large-scale industrial expansion in the twenty-first century. Processes were developed to create a new product: the multicrystalline silicon solar cell. Efficiencies were increased from 12% to 17% for monocrystalline silicon and 15% for multicrystalline. Photovoltaic modules were robust, with 20-year lifetime guarantees. Production processes moved from manual batch to highly automated. The rapid expansion that occurred from 2000 onward was based on the simple screen-printed wafer with silicon nitride front side ARC and passivation with an aluminium BSF. Solar cell designs capable of achieving much higher efficiencies were also demonstrated in this period. These advances and their gradual introduction into high-volume production will be described in Chapter 8.

References

1 International Technology Roadmap for Photovoltaics 6th edn (April 2015).
2 IRENA Solar Photovoltaics 1 (2012).
3 T.M. Bruton: Solar Energy Materials and Solar Cells 72 (2002) 3–10.

4 https://www.usgs.gov/centers/nmic/silicon-statistics-and-information accessed 7 August 2020.

5 PV Magazine, 20 August 2019.

6 M.A. Green: Progress in Photovoltaics 17 (2009) 183–189.

7 https://www.rsc.org/periodic-table/element/14/silicon accessed 7 August 2020.

8 M. Watt: Photovoltaics 1 (1993) ch. 1.

9 V. Fthenakis et al.: Progress in Photovoltaics 7 (1999) 489–497.

10 M.P. Bellman et al.: Proceedings of the 32nd EUPVSEC (2016) 1758–1763.

11 J. Zhao et al.: IEEE Transactions on Electron Devices 38 (1991) 1925–1934.

12 JEITA 2020 Production Forecast for Global Electronics and Information Technology Industries.

13 J.L. Gray in 'Handbook of Photovoltaics Science and Engineering' eds. A. Luque and S. Hegedus pub: Wiley (2003) 61–112.

14 M.A. Green: 'Silicon Solar Cells: Advanced Principles and Practice' pub: Bridge (1995).

15 M.A. Green: IEEE Transactions on Electron Devices 31 (1984) 671–678.

16 J.L. Gray: 'Handbook of Photovoltaics Science and Engineering' pub: Wiley (2003) 74–78.

17 K. Kimura: Proceedings of the International PVSEC1 (1984) 37.

18 T.M. Bruton et al.: Proceedings of the 10th EUPVSEC (1991) 667–669.

19 P. Woditsch et al.: Solar Energy Materials and Solar Cells 72 (2002) 11–26.

20 B. Ceccaroli and O. Lohne: 'Handbook of Photovoltaics Science and Engineering' pub: Wiley (2003) 179.

21 W.C. Sinke et al.: Proceedings of the 24th EUPVSEC (2009) 447–455.

22 W. Callaghan and R. McDonald: Flat Plate Solar Array Project Final Report (1984) DOE/JPL-1012-125.

23 J. Lindmayer and Z. Putney: Proceedings of the 15th IEEE PVSC (1981) 572–575.

24 M. Rodot in 'Advances in Solar Energy' ed. K.W. Boer pub: Springer (1983) 133–143.

25 B. Ceccaroli and R. Tronstad: 'Solar Silicon Processes' pub: CRC Press (2016) ch. 4.

26 P. Woditsch et al.: Proceedings of the E-MRS Spring Meeting Symposium E (2002) 11–26.

27 H. Aulich et al.: Proceedings of the International PVSEC1 (1984) P-11-1.

28 F. Slootman et al.: Annales de Chimie et de Physique 12 (1987) 401.

29 L.P. Hunt et al.: Proceedings of the 12th IEEE PVSC (1976) 125–129.

30 P. Pries et al.: Proceedings of the 27th EUPVSEC (2012) 1023–1025.

31 https://www.elkem.com/silicon-products/polysilicon accessed 7 August 2020.

32 R.H. Robinson et al.: Proceedings of the 14th IEEE PVSC (1980) 54–57.

33 W.C. Sinke et al.: Proceedings of the 24th EUPVSEC (2009) 447–455.

34 T.M. Bruton et al.: Proceedings of the 14th EUPVSEC (1997) 11–16.

35 M. Rogol: Proceedings of the 3rd Photon Solar Silicon Conference (2006) paper 01.

36 G. Holman: Proceedings of the 3rd Photon Solar Silicon Conference (2006) paper 03.

37 H. Oda: Proceedings of the 3rd Photon Solar Silicon Conference (2006) paper 05.

38 Y. Jiang et al.: Proceedings of the EPD Congress (2013) 173–183.

39 GTAT Press Release 11 June 2016.

40 https://www.energytrend.com/solar-price.html accessed 7 August 2020.

41 J. Perlin: 'From Space to Earth- the story of Solar Electricity' pub: Harvard University Press (2002) 166.

42 G.P. Willeke and A. Räuber: Semiconductors and Semimetals 87 (2012) ch. 12.

43 H. Fischer and W. Pschander: Proceedings of the 12th IEEE PVSC (1976) 86–92.

44 J. Lindmayer: Proceedings of the 14th IEEE PVSC (1980) 82–85.

45 F. Schmid in 'Solar Power for the World' ed. W. Palz pub: Pan Stanford (2014) 257–274.

46 K.A. Dumas et al.: Proceedings of the 15th IEEE PVSC (1981) 954–958.

47 T.M. Bruton et al.: Proceedings of the 22nd IEEE PVSC (1991) 1010–1014.

48 D.W. Cunningham: pers. comm.

49 https://www.csrwire.com/press_releases/22520-GT-Solar-to-Deliver-HEM-Furnaces-to-Deutsche-Solar accessed 7 August 2020.

50 Data Sheet DSS™850GTAT Corp (2016).

51 S. Qiao et al.: Proceedings of the 31st EUPVSEC (2015) 300–304.

52 US Patent 20070169684 18 June 2007.

53 V. Prajapati et al.: Proceedings of the 24th EUPVSEC (2009) 1171–1174.

54 K. Nakajima et al.: Proceedings of the 24th EUPVSEC (2009) 1020–1022.

55 Y. Yang et al.: Progress in Photovoltaics 23 (2015) 340–351.

56 P. Krenckel et al.: Proceedings of the 32nd EUPVSEC (2016) 289–293.

57 P. Varadi: 'Sun Above the Horizon' pub: Pan Stanford (2014) 89.

58 W. Koch: 'Handbook of Photovoltaics Science and Engineering' pub: Wiley (2003) 224–230.

59 J-I. Bye et al.: Proceedings of the 26th EUPVSEC (2011) 956–960.

60 Meyer Berger Fact Sheet DW288 Series 3 and DS271-264 (2016).

61 M.P. Bellman et al.: Proceedings of the 32nd EUPVSEC (2016) 1758–1763.

62 I. Tobias et al.: 'Handbook of Photovoltaics Science and Engineering' pub: Wiley (2003) 268–270.

63 A.D. Haigh: Proceedings of the 12th IEEE PVSC (1976) 360–361.

64 K. Kimura: Proceedings of the International PVSEC1 (1984) 37–41.

65 I. Tobias et al.: 'Handbook of Photovoltaics Science and Engineering' pub: Wiley (2003) 283–284.

66 T.M. Bruton and D.W. Cunningham: European Patent Application 89300843.3 25 October 1989.

67 A. Aberle and R. Hezel: Progress in Photovoltaics 5 (1997) 29–50.

68 W.J. Soppe et al.: Proceedings of the 16th EUPVSEC (2000) 1420–1423.

69 W.J. Soppe et al.: Progress in Photovoltaics 13 (2005) 551–569.

70 R.B. Godfrey et al.: Proceedings of the 16th IEEE PVSC (1982) 892–894.

71 D. Jordan: pers. comm.

72 V.G. Weizer and M.P. Godlevski: Proceedings of the 17th IEEE PVSC (1984) 134–136.

73 M.B. Spitzer et al.: IEEE Transactions on Electron Devices 31 (1985) 546–550.

74 K. Shiraswa et al.: Proceedings of the 14th EUPVSEC (1997) 384–387.

75 K. Shiraswa et al.: Proceedings of the 11th International PVSC (1993) 91.

76 J. Nijs et al.: Solar Energy Materials and Solar Cells 41/42 (1997) 101–107.

77 E. Kossen: Proceedings of the 25th EUPVSEC (2010) 2099–2104.

78 J. Donon et al.: Proceedings of the 17th IEEE PVSC (1984) 296–300.

79 T. Machida et al.: Proceedings of the 22nd IEEE PVSC (1991) 1033.

80 T. Saitoh et al.: Progress in Photovoltaics 1 (1993) 11–23.

81 R. Einhaus et al.: Proceedings of the 26th IEEE PVSC (1997) 167–170.

82 K. Shirasawa: Progress in Photovoltaics 10 (2001) 107–118.

83 I. Melnyk et al.: Proceedings of the 20th EUPVSEC (2005) 1403–1406.

84 J. Szlufcik et al.: Proceedings of the 14th EUPVSEC (1997) 380–383.

85 http://www.centrotherm.world/unternehmen/geschichte.html accessed 7 August 2020.

86 G.P. Willeke: Proceedings of the 29th IEEE PVSC (2002) 53–57.

87 K. Reinhardt and W. Kern: 'Handbook of Silicon Wafer Cleaning Technology' pub: Elsevier (2008).

88 http://www.appliedmaterials.com/products/applied-baccini-soft-line accessed 7 August 2020.

89 P. Schmitl et al.: Energia Procedia 27 (2012) 664–669.

90 G. Mathiak et al.: Proceedings of the 36th EUPVSEC (2019) 835–843.

91 http://www.mondragon-assembly.com/solar-automation-solutions/ accessed 7 August 2020.

92 J. Bennema: 'Solar Power for the World' ed. W Palz pub: Pan Stanford (2014) 341–345.

5

Evolution of Photovoltaic Systems

5.1 Introduction

Chapter 3 described the early years of solar installations, reporting on typical systems up to 1990. The world market was then dominated by small off-grid systems, both for domestic use such as lighting in remote housing – particularly in developing countries – and for professional systems such as telecommunications and coastal navigation, which had applications globally. Since the terrestrial market opened in the early 1980s, a number of factors have influenced its development. Initially, the technology was expensive, and only off-grid markets were cost-effective. Photovoltaics were preferred for professional systems not on its own cost basis, but for the costs it avoided in having to provide batteries to offshore installations or make helicopter trips to remote hilltop locations. Of course, the space satellite market was the ultimate off-grid application, where no further servicing could be done after launch. In the off-grid domestic market, the high initial cost of a photovoltaic system made purchasing difficult for the poorest people, and the market was dominated by government projects and charitable NGOs. These were relatively slow-growing markets. As described in Chapter 6, it was recognised that in order for photovoltaics to become an affordable technology, a much larger scale of manufacturing was needed, and various incentive programmes were thus put in place to stimulate market growth and, latterly, to meet carbon dioxide reduction targets. As the market grew, costs fell, and the size of photovoltaic systems increased. Meanwhile, new market segments were opened. This evolution is shown schematically in Figure 5.1.

While price remained high, only small-scale off-grid markets could be sustained. As prices fell, the domestic grid connected, commercial markets opened, and system size increased from typically 2 kWp up to 200 kWp. This represented domestic applications of 2–10 kWp and commercial systems (on corporate building and warehouses, carpark canopies, etc.) of 50–200 kWp. Additionally, there were ground-mounted systems of tens of kilowatts, particularly in agricultural sites in Germany. Above 200 kWp, systems tended to be ground-mounted standalone arrays. The largest are now over 1 GWp

Photovoltaics from Milliwatts to Gigawatts: Understanding Market and Technology Drivers toward Terawatts,
First Edition. Tim Bruton.
© 2021 John Wiley & Sons Ltd. Published 2021 by John Wiley & Sons Ltd.

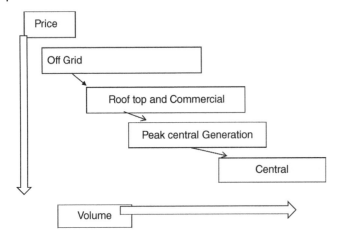

Figure 5.1 Schematic evolution of the photovoltaics market as price fell

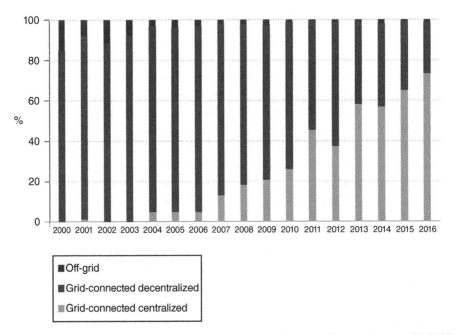

Figure 5.2 Evolution of the world photovoltaics market by sector (*Source:* Courtesy IEA PVPS)

in size [1]. As volume increased, manufacturing costs fell and new larger applications opened up, further increasing volume and driving down costs to today's historically very low level.

The shift in the market is illustrated in Figure 5.2. In 2000, the off-grid market was still significant in global terms, when the total market was around 200 MWp. It typically grew at 10% p.a., however, and so was rapidly overtaken by the grid-connected sector, where intermediate-scale domestic/commercial systems dominated. By 2007, the large centralised utility-scale plant was emerging from the demonstration market to become the dominant sector by 2015, as predicted by Figure 5.1. This chapter details the current status and likely development in these different market sectors, namely off-grid, decentralised on-grid, and utility-scale grid-connected.

5.2 The Off-Grid Market

While this is the smallest of the three current market sectors, at a social level it remains highly significant. The most recent studies indicate that the number of people living without electricity around the world may be as high as 1.3 billion (18% of the global population), 1 billion of whom live in areas with the best solar resources [2]. The potential for photovoltaics to remedy this has long been recognised. In 1990, Wolfgang Palz promoted the 'Power for the World' initiative, which proposed offering photovoltaics at 10 Wp per person to everyone living without electricity over the course of 20 years. This represented an annual production of 500 MWp at a time when the world market was perhaps 50 MWp, and thus seemed impracticable. However, the benefits in terms of improved water purification and supply for people, crops, and animals, improved lighting, health care, and communication, and novel economic activities were tantalising [3]. While 500 MWp p.a. was a huge extrapolation of the market in 1990, by 2019 it was about 0.5% of the global supply. In 2018, Palz expressed his frustration that so little had been done to aid the world's poorest through electrification [4].

Despite this negative view, progress has been made, although arguably not at the pace required. The World Bank has an active programme to provide electricity to rural communities in developing countries, and publishes a biannual report detailing its progress [5]. The market remains very large, with 840 million people globally without any access to electricity and a further 1 billion with unreliable grid connections (although making the distinction between no and an intermittent or poor connection is difficult). The majority of those with no access are in sub-Saharan Africa, while those with a poor grid connection are in South East Asia. The World Bank recognises three market segments: Pico, which is lanterns and other appliances up to 11 Wp; the Solar Home System (SHS), which would power lighting, phone charging, and television at above 11 Wp; and off-grid appliances, which includes a wide range of products powered by a DC supply. Global sales for all three sectors were estimated at $1.75 billion in 2019 [5], and it is anticipated that this will double by 2030, although many financial and commercial hurdles must first be overcome.

The competitiveness of off-grid systems is illustrated in Figure 5.3. The levelised cost of electricity (LCOE) for three location with insolation from good to average is compared with the LCOE for a standalone diesel generator, which is the main alternative to photovoltaics in remote locations. It can be seen that photovoltaics with batteries is cost-effective in most of the markets where electrification is needed, assuming an 8% discount rate. The challenge is in the initial investment cost, where a diesel generator

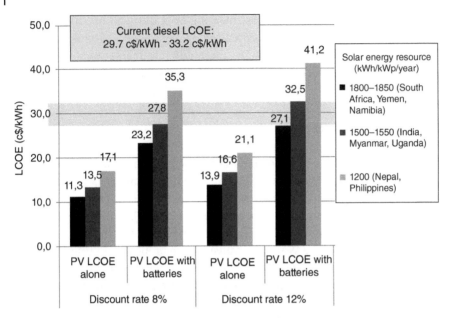

Figure 5.3 Levelised cost of electricity in different off-grid locations versus standalone diesel generator costs [2] (Courtesy EUPVSEC)

would cost around $300 compared to around $2100 for a solar system [2]. However, diesel has an ongoing cost of $250/year for fuel, plus issues of availability and maintenance of the system. Many of the SHSs are displacing kerosene lighting, which is also more expensive than photovoltaics and brings health hazards associated with fumes. Both alternatives are sources of CO_2, and meeting the estimated total 640 GWp demand by photovoltaics would save 1.55 Gt/year.

So, while photovoltaic systems cost less over the course of their lifetime, including the cost of replacement batteries, the two major obstacles have been the high initial cost to inherently poor users and the logistics of supplying the market and providing after-sales service to areas which are by definition remote from urban centres. Although these needs have been known about for 30 or more years, no single mechanism has been successful in meeting them. For grid-connected systems, mechanisms tested have included investment subsidies, feed-in-tariffs (FITs), and power purchase agreements (PPAs), the latter of which have ultimately proved the most successful for long-term sustainable growth in installations. For off-grid systems, a number of methods have been attempted. Simple aid projects, where solar systems are given to remote communities, have been unsuccessful and are now discouraged [5]. Several charities and NGOs have done good work. The UK charity Solar Aid has installed over 1.7 million solar lights in Africa (see Figure 5.4), sending education teams to remote locations to train end users, who purchase the lights through various finance arrangements. It reports improved health and better performance of children at school. Of growing interest is a new funding route termed 'PAY as YOU GO' (PAYGO). This is being

Figure 5.4 Solar lighting in a Zambian home (*Source:* Courtesy of Solar Aid/Corrie Wingate)

developed by private entrepreneurs, who raise capital and then provide SHSs at a low upfront cost and make monthly charges based on system use. Local infrastructure is used to supply and maintain the systems. Essential to success are a good mobile telephone system and mobile money, which make payment easy. The development of PAYGO has been strong in Central and West Africa, where this infrastructure exists. In India, where there is great potential, PAYGO is limited by the minimal mobile money facilities available. This system is still at an early stage of development, but $773 million was raised between 2012 and 2017. Annual growth rates of 140% have been achieved, and the World Bank sees it as the main vehicle for addressing the off-grid market, with sales potentially reaching $6–7 billion in 2022. PAYGO accounted for 24% of off-grid sales in 2019 [5].

The developing world is not the only place with significant off-grid markets. In the United States, 3700 houses were electrified off-grid in 2017. Australia is very sparsely populated outside its major cities, and many rural homes are off-grid. It also has a significant remote mining electricity demand, five times that of remote communities. Historically, this market has been heavily dependent on fossil fuels, particularly diesel. However, crude oil prices rose to over £100/barrel in 2013, stimulating users to look for more price-stable energy sources. The Australian government put in place a financial support mechanism (Rural Australia Renewables, RAR) to stimulate the take-up of renewable energy in remote areas and provide learning for future development [6]. The effect of this programme can be seen in Figure 5.5. The market was relatively stable prior to 2011, at less than 2 MWp p.a. The RAR introduced a significant growth in capacity, in addition to the effect of falling prices. Installations at mining sites are typically 10 MWp, while remote communities use 1 MWp systems in mini-grids. The RAR projects are still under evaluation for long-term stability of supply, but their feasibility has been demonstrated. It must be noted that at the same time, due to government stimulus in urban areas, 5 GWp of photovoltaics were installed in Australian cities.

While the number of people without electricity is a challenge, many so-called electrified communities do not enjoy a stable grid power supply, and massive efforts are

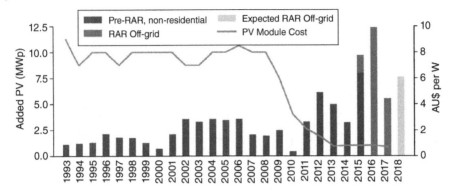

Figure 5.5 Annual photovoltaic capacity additions for off-grid nonresidential applications in Australia, with system cost data from [6] (*Source:* B. Herteleer et al: Proc 33rd EUPVSEC (2017) 2057-2066)

required to raise their power availability. The World Bank is addressing utility connection through its scaling programme, but the advantages of decentralised photovoltaic generation should not be ignored. These include rapid deployment, minimal dependence on expensive power transmission infrastructure, and easy integration with storage [2].

5.3 The Decentralised Grid-Connected Market

The off-grid market was dominant until the early 1990s, when a dramatic shift took place in developed countries toward the use of photovoltaic systems in already grid-connected areas. Germany and Japan became the leaders in this field. It was not immediately obvious why anyone should go to the expense of installing a photovoltaic system on a residential or commercial building when there was already an abundant power supply in place at around a third of the cost of the photovoltaic-generated power. However, a number of factors drove this development. The Chernobyl Nuclear Power Plant accident in 1986 created a widespread awareness of the dangers of nuclear power and gave strength to an already significant antinuclear lobby, particularly in Germany. For example, photovoltaics journalist and publisher Franz Alt became an ardent renewables advocate [7]. At the same time, there was growing awareness of the environmental impact of fossil fuel-based power generation, culminating in the UN Conference on Environment and Development in 1992, also known as the Rio de Janeiro Earth Summit. Both factors reinforced a widespread belief in the harmful impacts of global industrialisation and its consequences for the environment, particularly in generating CO_2 and hence producing global warming and climate change. This global recognition led to the Kyoto Protocol, in which 192 nations adopted binding targets for the reduction of

CO_2 emissions, particularly from electricity-generating power plants – although the United States was not one of them.

Walter Sandtner, author of the German '1000 Roof Programme', describes the development of grid-connected photovoltaics applications as having three phases [8].

5.3.1 The Research Phase: 1974–1989

Prior to 1989, grid-connected domestic-scale systems were purely proof-of-concept projects. The aim was to show that photovoltaics could be successfully deployed in a domestic dwelling and significantly contribute to its electricity consumption. A typical innovative house is shown in Figure 5.6. It was designed with good passive solar features, a 3.5 kWp photovoltaic array, and solar thermal water heating. It cost the same to construct as a conventional house, and its CO_2 production was just 148 kg p.a., as opposed to the typical 6500 kg [9]. Many similar one-off conceptual designs were built around the world.

5.3.2 The Demonstration Phase: 1989–2000

In 1987, there were only six grid-connected photovoltaic houses in Germany [8]. In the United States, the Gardner demonstration site was used to identify issues of grid connection [10], while in Japan, the problems of multiple small photovoltaic systems feeding into the grid were characterised [11]. There was an active debate over the benefits of having many small photovoltaic systems distributed at the point of use, rather than

Figure 5.6   Eco House Oxford (*Source:* Courtesy of BP Archive/Susan Roaf)

the large central plants that had been the norm for electricity production by fossil- and nuclear-powered sources [11]. The advantages were seen as reducing transmission losses (typically 30%) in well-maintained lines, saving the use of additional land (important in countries like Japan and Switzerland), and cutting costs through the use of existing building infrastructures. The first working party in Germany rejected the possibility of connecting small photovoltaic systems to the grid because of the lack of standards, but as pressure mounted the programme was agreed to [8]. In Japan, the 1993 Guideline of the Technical Requirement for PV Grid Connection with Reverse Flow was agreed to after similar debate and the deployment of residential photovoltaic systems began [12].

In Germany, the 1000 Roof Programme was an immediate success. Each of the 16 German Lander were allocated a number of systems. A 70% subsidy was applied to the system price, with 50% coming from the federal government and 20% from the Lander. One Lander refused to apply its subsidy and was undersubscribed, while all the others were oversubscribed. Such was the interest that, ultimately, over 2000 roofs were installed for a total of 6 MWp capacity. When the programme was broadened to allow non-German manufacturers to supply panels, the system price fell by 30% and considerable experience was gained. The schematic of a typical system is shown in Figure 5.7. The components are a photovoltaic array (typically 2–3 kWp for a domestic system), an electrical DC power into AC power inverter, and a line feeding into the local grid. The introduction of the inverter was the most significant change in moving from small off-grid DC systems to on-grid AC systems. The modules were usually roof-mounted, with a low profile close to the slope of the existing roof (see Figure 5.6), which in Northern Europe was fairly optimal.

The systems were closely monitored in order to determine the actual useful electrical output from the system [13]. Prior to the 1000 Roof Programme, installed capacity in terms of kWp (DC) had been the main parameter characterising photovoltaic systems. Now, the focus changed to measuring actual output using the annual final yield

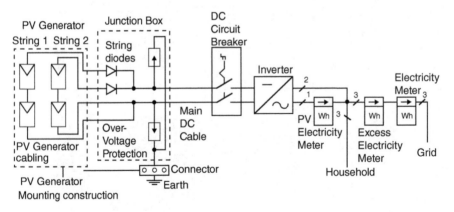

Figure 5.7 Block diagram of a typical 1000 Roof Programme domestic system [13]
Source: B. Decker and U. Jahn: Solar Energy 59 (1997) 127-133

Y_f, defined as the ratio of the annual photovoltaic energy output divided by the power of the system at STC and given in units of kWh/kWp (STC) [13]. In the earliest 1000 Roof systems, there was quite a range of Y_f; for example, it varied from 430 to 875 kWh/kWp in 1994. This reflected the demonstration nature of the programme at this point in time. Clearly, many lessons were to be learnt in terms of the quality of the installation, the operation of the inverter, and the siting of the systems. An alternative measure of system performance was the performance ratio (PR), defined as the ratio between actual photovoltaic energy output and theoretically available energy. PR varied from 0.4 to 0.85 for systems installed between 1990 and 1994, with a mean of 0.65; this increased to a mean of 0.75 for systems installed between 1996 and 2002, with many reaching over 0.80 [14]. Inverter failure in the early years was a major contributor to loss of output.

The 1000 Roof Programme ended in 1994 with no federal programme to replace it. This caused a major decline in German manufacturing, with leading companies closing production and acquiring capacity in the United States. Siemens Solar purchased ARCO Solar and ASE purchased the EFG technology from Mobil Solar, while production at AEG Telefunken ceased [15]. However, as discussed in Chapter 6, local initiatives promoted the use of photovoltaics in homes by mandating regional utilities to purchase electricity from domestic systems at the rate of 1.8 DM/kWhr (~€0.9/kWhr). There was strong popular support for this scheme, and it quickly spread from Aachen, where it was first introduced, to 40 towns and cities across Germany [16]. The impact of these measures can be seen in the growth of the German market, as shown in Figure 5.8. This led to general acceptance of the grid-connected photovoltaics concept and created credibility for the next proposed federal programme, the 100 000 Roof Programme – which in fact was overtaken by the German Renewable Energy Act, which then stimulated a GWp market, as described in Chapter 6.

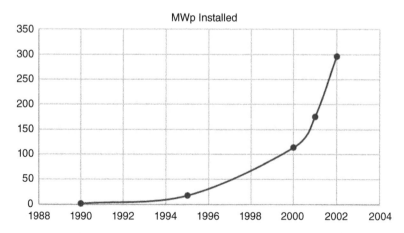

Figure 5.8 Cumulative installations in the German market from 1990 to 2002, in MWp [17] (*Source:* IEA PVPS report for PV Power Applications in Germany 2016)

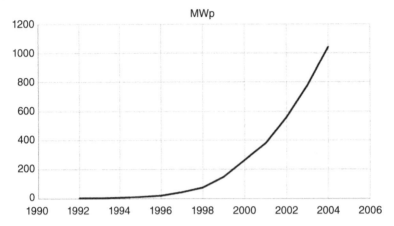

Figure 5.9 Growth of the Japanese distributed grid-connected market from 1992 to 2004, in terms of cumulative installed capacity in MWp per year [19] (*Source:* IEA PVPS Report for PV Power applications in Japan 2017)

While Germany took an early lead, Japan was quick to follow [15]. Its first programme, in 1992, was the field test Project for Public Facilities. This was followed by the Residential PV System Monitor Project in 1994 and the Residential PV Dissemination project in 1997. The market was stimulated by a combination of capital grants for system purchases and a net metering scheme for purchases of surplus electricity [18]. These measures helped to establish a strong Japanese manufacturing base, from which photovoltaic products were also exported widely. The growth of the market for domestic systems is shown in Figure 5.9.

By 2003, Japan had become the largest annual photovoltaics market by some margin, with twice the installed base of Germany and the largest manufacturing capacity in the world. However, things changed quite dramatically in 2004, when Germany overtook Japan again with annual installations of 565 MWp and the German market expanded into its full commercialisation phase.

5.3.3 The Commercial Phase: 2000–2019

As can be seen from Figures 5.8 and 5.9, the market evolved relatively slowly up to 2000, but growth was explosive from that point onward and photovoltaics became established as a significant global energy source. The world market first passed 100 MWp in annual shipments in 1997 [20], but just seven years later it delivered 1 GWp of installations globally. The market at that time was led by Germany, with a number of other European nations following suit (notably Spain and Italy). The driving force in Germany was the FIT scheme, enabled by the Renewable Energy Law. The principle was that all electricity consumers paid a little more to allow payments to photovoltaic system owners, who could then make an economic return on their investment. The market was divided into three segments with appropriate FITs, namely

Figure 5.10 Large domestic grid-connected system in Germany (Courtesy BP Archive)

domestic (1–5 kWp), commercial and agricultural (50–200 kWp), and large free-standing (0.5–50 MWp) [15]. All three enjoyed substantial support. A typical large domestic system is shown in Figure 5.10, while a large commercial array integrated into an industrial facade is shown in Figure 5.11.

There was a considerable imbalance between production capacity and market demand in Europe, however. Figure 5.12 shows the flow of modules into the European market in 2006, revealing a nearly 1 GWp deficit, filled by imports from Japan and elsewhere in Asia. The United States was also a net importer, albeit with a smaller market. This highlights a trend that was to be even more dramatically realised after 2010. The stimulation of the market in Germany resulted in a good climate for investment in photovoltaic manufacturing and additional capacity was built. SolarWorld expanded its activities, acquiring first the assets of Bayer Solar and then Shell Solar to become one of the world's largest companies. RWE initiated a round of expansion at its Alzenau plant, renaming the NUKEM/ASE operation to RWE Solar, with both conventional solar cell manufacture and also manufacture of the EFG technology it had acquired from Mobil Solar [15]. The Q cells company was founded with private funding, and by 2009 it had become the largest manufacturer in Germany at 518 MWp and, briefly, the world. It also diversified into cadmium telluride and copper indium diselenide (CIGS) manufacture, although at a small scale. In 2004, of the world's ten largest manufacturers, four were Japanese and three were German [20]. While domestic production increased, Europe, and particularly Germany, represented a very attractive export market, with good economies and long-term government

Figure 5.11 Large solar system integrated into the facade of Stadtwerke, ULM, Germany (*Source:* Courtesy of BP Archive)

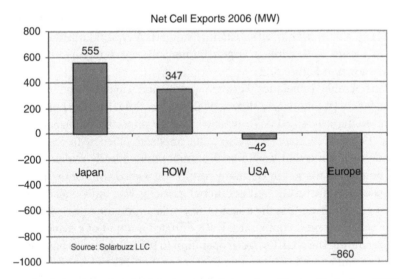

Figure 5.12 Flow of modules into main global markets (*Source:* Marketbuzz 2004 Market review: Publisher Solarbuzz LLC (now IHS Markit))

commitments to the funding of photovoltaics. There was little change in the technology being deployed. The silicon solar cell on a monocrystalline or multicrystalline wafer with full aluminium back surface field (BSF) was the dominant low-cost product. The manufacturing technology had matured at this point in time, so that there was there were several companies in both Europe and the United States that were able to offer

Table 5.1 Annual installations and cumulative installed capacity by country in 2011 [21]

Country	Capacity connected in 2011 (MWp)	Cumulative installed capacity to 2011
Italy	9000	12 500
Germany	7500	24 700
China	2000	2900
USA	1600	4200
France	1500	2500
Japan	1100	4700
Australia	700	1200
United Kingdom	700	750
Belgium	550	1500
Spain	400	4200
Greece	350	550
Slovakia	350	500
Canada	300	500
India	300	450
Ukraine	140	140
Rest of world	1160	6060
Total	**27 650**	**67 350**

(*Source:* Modified from Global PV market report 2012 published by the European Photovoltaic Industry Association)

high-quality turnkey silicon photovoltaic production facilities. This stimulated investment and the growth of manufacture in Asia, particularly in China and Taiwan. By 2007, there were four Chinese companies in the top ten manufacturers, and by 2011 there was only one non-Asian one. The expansion in China was aided by easy access to proven manufacturing equipment and by good R&D support available internationally. China's first company, Suntech Power, was founded in 2001 by a former researcher from the renowned UNSW photovoltaics research team, while the late Prof. Stuart Wenham became its Chief Technology Officer.

Although a shift in the centre of manufacturing was occurring in 2011, Europe remained the dominant market. Table 5.1 summarises the annual and cumulative installations for that year [21]. It can be seen that 73% of annual global installations were in Europe, as was 70% of total installed global photovoltaic capacity. As shown in Figure 5.2, 60% of the market was still decentralised grid-connected, although the shift to central generation was increasingly evident. The European market began to grow much more slowly from 2011 onward. System prices had fallen strongly and there was a need to shift to self-consumption of photovoltaic electricity rather than export of surplus electricity to the grid. FITs were adapted to reflect this, particularly in Germany [23], and stimulus was given to install domestic battery storage systems, and

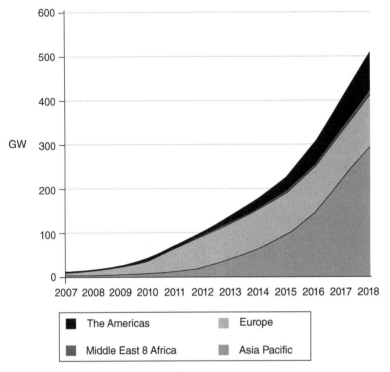

Figure 5.13 Evolution of cumulative installed photovoltaic capacity by region [22]
(*Source:* IEA-PVPS Trends in Photovoltaic Applications 2019)

10 000 were installed in Germany in 2013. The evolution of cumulative installed capacity is shown in Figure 5.13, which reveals the rapid growth in the Asian and US markets compared to Europe. However, the European market remains significant, with 11.3 GWp installed in 2018 [24].

While the utility-scale large centralised pant has become the largest market sector, as discussed in the next section, total rooftop installations still were 29.7 GWp in 2018, compared to 72.7 GWp for the ground-mounted utility sector [24]. What is the future for distributed on-grid photovoltaics? The achievement of 'grid parity' (i.e. when the consumer could install a photovoltaic system with an LCOE below the retail electricity rate) was seen historically as the point when photovoltaics became commercially viable. As discussed later, this was achieved for most countries in the early 2010s. However, installation of distributed photovoltaic systems is a function of a complex interaction of initial investment, rates for exported electricity, and other subsidy mechanisms. It is likely going forward that the major growth will come from systems where there is substantial self-consumption. A good example is a supermarket with high daytime electricity loads for lighting, refrigeration, and air-conditioning plus a large, generally flat roof space, as shown in Figure 5.14. The commercial sector for distributed grid-connected generation is likely to be a highly significant long-term

Figure 5.14 Large-scale solar system on a supermarket with high self-consumption (*Source:* SEIA)

market despite the growth in utility-scale plants described later. Many major corporations such as Google aspire to become carbon-neutral. In 2019, through direct purchases and PPAs, 19.5 GWp was purchased by corporations globally – a 40% increase on 2018 [25].

The rise of electric vehicles and their charging requirement, plus the ability to move charge to and from a vehicle, will also see the distributed photovoltaic sector continue to be significant, especially in the context of the growth of smart cities [24].

5.3.3.1 Achievement of Grid Parity

As shown in Figure 5.1, it was assumed historically that the rooftop market would be the first grid-connected market to become cost-effective. It had always been accepted that photovoltaic systems could reliably deliver domestic electricity. The issue was, could they do so cost-effectively? Back in 1956, Chapin had calculated that it would cost a home owner $1.43 million to fully electrify their house with photovoltaics [26]. As discussed in Chapter 2, the immediate challenge of the 1973 energy crisis was how to make photovoltaics affordable. It turned out to be a long journey, as it was only when the economies of scale from gigawatt manufacture were achieved that the cost of photovoltaic modules fell to the point where grid parity could be achieved.

Similar savings were needed in inverter costs and in standardising installation and grid connections. These came in the same way, as the economies of scale kicked in for the manufacture of these items. Inverter costs fell from around €1/W before 2010 to less than €0.1/W today [27]. A detailed study was conducted in 2016 to calculate the

LCOE for different European countries [28]. It assumed a 30-year system life and calculated capital cost, maintenance costs, system degradation, rates of self-consumption, and retail electricity rates. The overwhelming conclusion was that for virtually all system sizes and locations, grid parity was achieved in European markets in 2016, with the possible exceptions of Sweden and Finland where retail electricity rates were low and the solar resource was also relatively low. Figure 5.15a shows the projected cost decrease for a 5 kWp rooftop system in Amsterdam, a medium European case. It can be seen that cost-effectiveness is determined by the value of the self-consumption of electricity and the weighted average cost of capital (WACC). For Amsterdam, this is achieved with a self-consumption of around 50% and a WACC of 2%. The value chosen for WACC is the biggest variable, and essentially is not determined by photovoltaic technology. Increasing the WACC in Amsterdam from 0 to 6% increases the calculated LCOE by over 80%. For a traditional calculation of grid parity, 100% self-consumption would be assumed [28]. The contrast with Rome (Figure 5.15b) shows the impact of the solar resource, with LCOE competitive in all WACC and self-consumption scenarios.

The achievement of grid parity in industrialised countries can be said to have taken 30 years from the start of a significant terrestrial photovoltaics market in the 1980s. The next challenge was perceived to be the generation or fossil fuel parity. In the early 2010s, the cost of solar modules and related costs were falling so rapidly that for many high-solar-resource countries photovoltaic electricity was not only available at competitive rates compared to fossil-fuelled power and other renewables but was actually the cheapest form of electricity generation.

5.3.3.2 Resolution of the Silicon Feedstock Supply

While crystalline silicon has been the dominant photovoltaics manufacturing technology, it is crucially dependent on a supply of hyper-pure silicon with total metallic impurities present only at the parts per billion level (semiconductor-grade polysilicon, sc-Si). Historically, this came from the suppliers to the global semiconductor industry, which was dominated by six companies that enjoyed a close long-term relationship with silicon wafer manufacturers. There was a great reluctance among these companies to recognise the growth potential of photovoltaics, or the market opportunity it represented [15]. Perhaps some of this reluctance came from the price demanded by the photovoltaics industry: the typical cost of sc-Si through the early years was $50/kg [29], but it was proposed that a price below $16/kg was essential to achieve low-cost photovoltaic modules [30]. Presumably, given the uncertainty over the ultimate photovoltaics market size, this did not represent a promising business opportunity to the mainstream silicon producers. Both Wacker and Union Carbide did receive government funding to produce low-cost polysilicon, but not by the traditional Siemens process [15].

The early silicon photovoltaics industry survived by using the scrap fall-out material from the semiconductor industry. This consisted of tops and tails of boules,

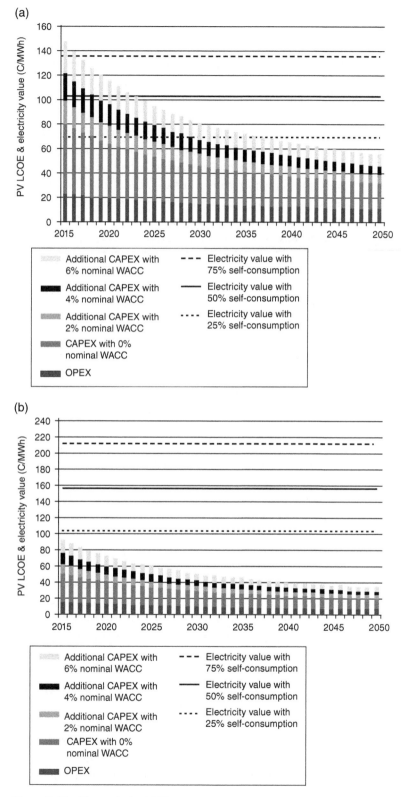

Figure 5.15 Photovoltaics LCOE and value of electricity for a residential 5 kWp roof-mounted system (a) in Amsterdam *Source:* E Vartiainen, G Masson and Ch. Breyer: Proc 32nd EUPVSEC (2016) pp 2836-2846 and (b) in Rome [28] (Courtesy EUPVSEC)

off-specification feedstock (either electronically or dimensionally), and reject and part-processed wafers. Around 10% of semiconductor feedstock fell into this category. In 2000, global polysilicon production was 20 000 tonnes p.a. [29], so that around 2000 tonnes p.a. was available for photovoltaics production. Given a worst-case scenario in which photovoltaic silicon required 15 tonnes of polysilicon per megawatt of solar cells, this made an annual production of 130 MWp possible. This remained the situation up until 2000. However, periodic shortages occurred, when surges in demand on the part of the semiconductor industry resulted in shortages of polysilicon, with materials previously considered scrap held back for its own use. This drove the photovoltaics industry to use whatever was available, including the highly doped silicon pot scrap material left in the crucible after Czochralski (CZ) growth, which required very labour-intensive manual sorting.

Despite many attempts on the part of the photovoltaics industry to engage in dialogue with the established producers to secure increased production and lower prices, no real engagement occurred. The response among polysilicon producers was broadly that the price was the price and it was the responsibility of the photovoltaics industry to be more efficient in its use of silicon in order to effectively increase available feedstock volume. This view was still being expressed, albeit implicitly, as late as 2009 [31].

It proved difficult to identify the true manufacturing costs of polysilicon, as production usually took place in large integrated chemical plants where biproduct streams were used for other, non-semiconductor products [31]. A reliable industry estimate was made as part of a MUSIC FM study, which concluded that a polysilicon price of €20/kg was sustainable assuming some minor relaxation of the product specification [32].

A crisis situation developed between 2000 and 2005 when the photovoltaics industry expanded far more rapidly than the supply of polysilicon, resulting in a massive increase in the spot price for polysilicon, as shown in Figure 5.16. Photovoltaics manufacturers had been paying around $20/kg for the semiconductor fall-out material. With the increase in demand, they had to compete for prime material with the semiconductor industry. It was reported that in some cases, spot prices above $300/kg were paid at the peak of the shortage [33].

Clearly, this situation was not sustainable. The lack of supply opened opportunities for other technologies, and First Solar in particular was able to enter the market with its cadmium telluride thin-film modules. There was an upsurge of interest in concentrating photovoltaic systems, and a number of existing companies like Amonix expanded their activities, while new entrants such as Concentrix and SolFocus enjoyed good levels of investment. Inevitably, the high prices paid for polysilicon, together with the strong growth prospects, also made this a strongly attractive business opportunity. At the same time, while expertise in polysilicon manufacture had previously been a closely guarded secret kept within the established industry, companies like GT Advanced Technologies were suddenly offering turnkey polysilicon production units. From 2007 onward, GT Advanced supplied 400 units, which were claimed to account for 30% of global polysilicon production [34]. A number of photovoltaic manufacturers sought to

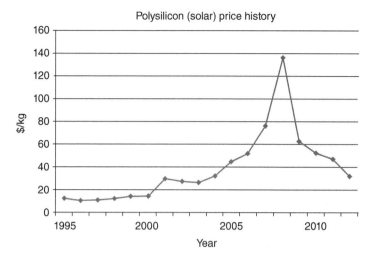

Figure 5.16 Price of polysilicon to the photovoltaics industry, 1995 to 2012 [15] *Source:* G. P. Willeke and A. Räuber: Semimetals and Seminconductors 87 (2012) chapter 3

become integrated manufacturers with their own polysilicon supply; for example, PV Crystalox erected an 1800 tonnes/year polysilicon plant, while other companies such as GCL, LDK, and Renesolar in China, M Setek in Japan, and OCI in Korea [15] all began their own production. In 2005, the Norwegian REC purchased the ASiMi polysilicon plant (formerly the Union Carbide plant) in the United States. The market reversed into oversupply and prices fell to below $30/kg by 2012, and European polysilicon plants were closed down. Since 2012, the supply of silicon has been stable, with the majority of that in China going to Chinese photovoltaics manufacturers. The traditional producers did expand production, albeit relatively slowly. Polysilicon prices are current low in historic terms, with prices for prime solar-grade material below $10/kg, indicating a more than adequate supply to the market [35].

5.4 Utility-Scale Grid-Connected Photovoltaic Systems

The title of this book is *Photovoltaics from Milliwatts to Gigawatts*, and Chapter 3 described the emergence of the milliwatt market for indoor consumer photovoltaic products. This section now looks at the emergence of the gigawatt-scale central grid-connected electricity-generating plant. As already described, as early as 1983 it was thought that for photovoltaics to be significant it would have to produce electricity in large utility-scale plants, but there is now a strong belief that the appropriate use for photovoltaics is in distributed generation, whether connected to a grid or not. The view is that the solar resource is available effectively everywhere and end users are also widespread, so there is no need for a large distribution infrastructure, with all its costs and associated losses. Nevertheless, there are also strong arguments for central generation plants. These can be summarised as follows:

1) Large utility plants are the fastest way to add photovoltaic power to the grid. It is simpler to ground mount a single 100 MWp plant than to convince 3.3 million consumers to install 3 kWp of roof-mounted systems each. This is the easiest and quickest route for a country to take in order to reduce its greenhouse gas emissions.
2) Large ground-mounted systems are much cheaper to install that rooftop arrays. In strong economies like the United States and Germany, large ground-mounted arrays typically cost 50% less than 5 kWp rooftop systems, while in lower-wage economies they cost 30–40% less [22].
3) In many emerging economies, while there may be large, rural unelectrified areas, the grid is often weak even in urban areas and adding photovoltaic capacity to existing grids greatly improves the availability of power. A good example is South Africa [36].
4) In terms of delivered electricity in good solar regimes, photovoltaics gives the lowest LCOE for new-build power generation (see later).
5) The lead time for the construction of a utility-scale photovoltaic plant is much shorter than that for a conventional fossil fuel or nuclear plant. Moreover, a photovoltaic plant can be commissioned in a staged process.

As can be seen from Figure 5.2, around 70% of installations in 2016 were utility-scale grid-connected photovoltaic systems. The reasons for this vary from country to country and with different models of funding. The three largest photovoltaics markets in 2017 were China, India, and the United States [24]. China and India have pursued aggressive photovoltaic deployment plans. An initial Chinese target of 105 GWp installed by 2020 was actually exceeded in 2017 when 52.8 GWp were installed in a single year, to give a total capacity of 130.8 GWp, and the Chinese incentive programme was adjusted in 2018 to slow growth. In India, the target was 100 GWp by 2020, 40% of which would come from rooftop sources. In total, 9.6 GWp were installed in 2017, but the rooftop programme was seriously behind schedule [24]. In the United States, the prime driving force is the Renewable Portfolio standard, which operates at state level and mandates electricity suppliers to purchase a certain amount of renewable electricity. Currently, 29 states have such policies, and a number are proposing moving the requirement to 50% renewable. Total US installations were 10.7 GWp in 2018, mainly in utility-scale plants [22].

It is reported that photovoltaic power accounts for the majority of new generating capacity installed worldwide [24]. Part of the reason for this is that it is the cheapest option. A study by the US bank Lazard [37] found that in a good-solar-resource region such as the US South West, the LCOE for new-build utility-scale photovoltaics was lower than that of any competing technology, including combined cycle gas turbine, coal, and nuclear, as illustrated in Table 5.2. If correct, this should be reflected in the latest electricity prices for tenders and PPAs for large power plants. In February 2018, a tender for 300 MWp in Saudi Arabia was won by a local company with a record low price

Table 5.2 Electricity generation costs for photovoltaics and other technologies [36]

LCOE $/MWh	Generating technology
40	Utility-scale solar photovoltaics (silicon technology)
41	Wind
56	Gas combined cycle
91	Geothermal
109	Coal
141	Solar thermal tower
155	Nuclear
175	Gas peaking

(*Source:* Modifed from Lazard's Levelised cost of Energy Analysis Version 13.0 November 2019)

of $0.0234/kWhr [24]. The lowest bid was actually below 2US cents/kWhr, but it was disallowed because it included bifacial modules, for which performance standards are not yet defined. The level of the successful bid is not a surprise as it reflects a long-term trend of decreasing prices, as described in Chapter 6. The lowest global tender is $0.01567 in Qatar, an area with low land costs and a high solar regime [38]. It is worthwhile noting here that the lowest contract price in Germany to date is $0.0355/kWhr [39].

As the utility-scale photovoltaic sector has grown, not only have costs decreased significantly but the largest plant sizes have grown. The first plants greater than 100 MWp were installed in 2011 (Mesquite, United States, 150 MWp; Sentftenberg, Germany, 168 MWp; Golmud, China, 200 MWp) [1]. By 2014, the largest plant built was the 550 MWp Topaz Solar Farm in the United States, using First Solar cadmium telluride modules, as shown in Figure 5.17. This is now just the 17th largest plant in the

Figure 5.17 550 MWp Topaz Solar Farm, United States (Courtesy First Solar)

world [40]. There are currently seven plants commissioned globally with capacity greater than 1 GWp, two of which exceed 2 GWp. The largest photovoltaic plant operating today is Bhadla in India, which was connected in March 2020 and has a capacity of 2.245 GWp [41].

It can be anticipated that the utility-scale photovoltaic plant will continue to grow year on year, meeting the needs of emerging economies with a power generation deficit but good solar resources. Utility-scale photovoltaics will provide the fastest route to the terawatt-scale installation that will be required by 2050.

5.5 Novel Applications

In 2015–16, pilots Bertrand Piccard and Andres Borschberg flew the photovoltaic-powered solar impulse aircraft on a 40000 km circumnavigation of the globe, including five consecutive days and nights crossing the Pacific Ocean. No fuel other than sunlight was used [42]. While this was an inspiring and epic undertaking, it appeared to have no commercial application. However, in February 2019, the UK mapping agency the Ordnance Survey announced it would be using a solar-powered unmanned aircraft, flying continuously for 90 days at 67000 feet, to carry out aerial surveillance (Figure 5.18). This was in preference to using satellites, which orbit too high to provide sufficient resolution for map making [43]. This illustrates the as yet untapped potential of photovoltaic power to address new market opportunities and applications as the need arises.

Figure 5.18 Unmanned ground surveillance drone capable of 90 days' continuous flight (*Source:* Courtesy of Ordnance Survey)

5.6   Summary

This chapter has described how photovoltaic power has matured into a global technology with profound social impacts in off-grid applications in improving health and economic opportunities. The technology has achieved cost-competitiveness against other types of energy source and is providing greenhouse gas-free power in both utility-scale and decentralised grid-connected power plants. Market saturation is some way away. To date, 512 GWp has been installed globally. With annual installations around 100 GWp, it is reasonable to expect a total cumulative capacity of 1 TWp by 2022 [24].

References

1 http://www.pvresources.com/en/pvpowerplants/top50pv.php accessed 7 August 2020.
2 H.J. Yu: Proceedings of the 35th EUPVSEC (2018) 7DV2.3.
3 W. Palz: 'Solar Power for the World' pub: Pan Stanford (2014) 155.
4 W. Palz: 'The Triumph of the Sun' pub: Pan Stanford (2018) 169.
5 Off-Grid Solar Market Trends Report January 2020.
6 B. Herteleer et al.: Proceedings of the 33rd EUPVSEC (2017) 2057–2066.
7 F. Alt in 'Solar Power for the World' ed. W. Palz pub: Pan Stanford (2014) 301–305.
8 W. Sandtner in 'Solar Power for the World' ed. W. Palz pub: Pan Stanford (2014) 335–339.
9 S. Roaf: 'Ecohouse2 – A Design Guide' pub: Architectural Press (2003).
10 E.C. Kern et al.: Proceedings of the 19th IEEE PVSC (1987) 1007–1011.
11 K. Takigawa et al.: Proceedings of the 19th IEEE PVSC (1987) 1001–1006.
12 O. Ikki and I. Kaizuka in 'Solar Power for the World' ed. W. Palz pub: Pan Stanford (2014) 418.
13 B. Decker and U. Jahn: Solar Energy 59 (1997) 127–133.
14 U. Jahn and W. Nasse: Progress in Photovolatics 12 (2004) 441–448.
15 G.P. Willeke and A. Räuber: Semimetals and Semiconductors 87 (2012) ch. 3.
16 H. Scheer in 'Solar Power for the World' ed. W. Palz pub: Pan Stanford (2014) 295.
17 IEA PVPS Report for PV Power Applications in Germany 2016.
18 O. Ikki and I. Kaizuka in 'Solar Power for the World' ed. W. Palz pub: Pan Stanford (2014) 419.
19 IEA PVPS Report for PV Power Applications in Japan 2017.
20 Marketbuzz 2004 Market Review pub: Solarbuzz LLC (now IHS Markit).
21 Global PV Market Report 2012 pub: European Photovoltaic Industry Association.
22 IEA-PVPS Trends in Photovoltaic Applications 2019.
23 IEA-PVPS Trends in Photovoltaic Applications 2014 26.
24 Global Market Outlook for Solar Power 2019–2023 pub: Solar Power Europe.

25 BloombergNEF 28 January 2020.

26 J. Perlin: 'From Space to Earth – The Story of Solar Electricity' pub: Harvard University Press (2002) 36.

27 Photon International October 2018 49.

28 E. Vartiainen, G. Masson, and C. Breyer: Proceedings of the 32nd EUPVSEC (2016) 2836–2846.

29 B. Ceccaroli and O. Lohne in 'Handbook of Photovoltaic Science and Engineering' eds A. Luque and S. Hegedus pub: Wiley (2008) 153–208.

30 J. Stone et al.: Proceedings of the 17th IEEE PVSC (1984) 1178–1183.

31 K. Hesse et al.: Proceedings of the 24th EUPVSEC (2009) 883–885.

32 T.M. Bruton et al.: Proceedings of the 14th EUPVSEC Barcelona (1997) 11–16.

33 https://www.icis.com/explore/resources/news/2011/12/28/9519292/outlook-12-asia-polysilicon-likely-to-be-soft-on-economy-trade accessed 7 August 2020.

34 https://gtat.com/about/ accessed 7 August 2020.

35 http://pvinsights.com/ accessed 7 August 2020.

36 T. Bischof-Niemz and K.T. Roro: Proceedings of the 31st EUPVSEC (2015) 3141–3260.

37 Lazard's Levelised Cost of Energy Analysis Version 13.0 November 2019.

38 Taiyang News 27 January 2020.

39 Taiyang News 24 February 2020.

40 https://en.wikipedia.org/wiki/List_of_photovoltaic_power_stations accessed 7 August 2020.

41 https://mercomindia.com/world-largest-solar-park-bhadla/ accessed 7 August 2020.

42 https://aroundtheworld.solarimpulse.com/adventure accessed 7 August 2020.

43 www.bbc.co.uk/news/technology-47196898 accessed 7 August 2020.

6

History of Incentives for Photovoltaics

6.1 The Chicken and Egg Problem

The challenge in 1973, when photovoltaics was first considered as a possible signifi-cant terrestrial energy source, was how to reduce the cost to a level where it could compete with fossil fuels and nuclear generation. As well as lowering the cost of raw materials and improving the manufacturing process, increasing the volume of manufacture was also seen as an important element. The question to be faced was how to migrate from the limited off-grid applications to a mass market. In the 1980s, the commercial market was entirely off-grid, with a few on-grid demonstrations. Although potentially large at a GWp scale, the off-grid market only grew very slowly. It was difficult to see how growth could be accelerated to achieve the volumes nec-essary to achieve the <$1/Wp module price for large-scale deployment [1]. Early studies had suggested that manufacturing costs could fall to below $1/Wp if 100 MWp annual manufacture in a single plant occurred [2]. However, this volume of production was more than ten times the world installations at that time, and the processes themselves were still in their infancy. A later study in 1997 based on actual manufacturing experience at the tens of MWp p.a. projected cost below €1/ Wp at 500 MWp p.a. manufacturing [3], although again this was more than five times the global market at the time. More recently, in 2009, a study was done in which 1 GWp p.a. manufacturing scale was cost modelled and a range of silicon cell technologies had module production costs close to €1/Wp [4]. In recent years, these paper studies have been overtaken by the reality of large production plants in East Asia capable of manufacture in the 5–10 GWp scale and module prices have fallen to below €0.5/Wp, or as low as €0.2/Wp [5]. This has vindicated the concept that economies of scale in manufacturing will produce the required cost reduction for photovoltaics to be a globally significant energy source.

That this level of production was achieved is in part due to the various national incen-tive programmes that were put in place. It was evident to decision makers as far back as the 1980s that the photovoltaics market would not grow organically at a sufficiently fast

Photovoltaics from Milliwatts to Gigawatts: Understanding Market and Technology Drivers toward Terawatts, First Edition. Tim Bruton.

rate to enable low cost potential to be attained [6]. The incentives offered can be divided into three types, corresponding roughly to their historical introduction – although there were overlaps, and different countries emphasised different routes:

1) *1980–1995*. Capital subsidises on system purchase.
2) *1995–2012*. Feed-in-tariffs (FIT).
3) *2012–present*. Tenders, auctions, and private purchase agreements.

6.2   Capital Subsidies on System Purchase

When the industry was very new and system cost was a major hurdle, it seemed the obvious thing to do was to use state intervention to reduce the upfront cost to consumers. One of the first programmes was implemented in California in 1976, when 10% credit was offered for single-occupancy homes, rising to 55% for larger homes and commercial properties. The US federal programme started in 1978 with 30% tax credit for the first $2000 of investment and 20% on additional investment up to $8000. In 1980, this was modified to 40% credit on a maximum $10000 investment [7]. There was a requirement that the system life should be 5 years, which now seems strange but was realistic at the time, as discussed in Chapter 2. However, all tax credits came to an end in 1985. This had a marked effect on the growth of the solar business, much of which had been in the United States up to that point. As shown in Figure 6.1, strong growth had occurred to 1985 at a compound rate of 26%, but between 1985 and 1986 there was virtually none, which shows the importance of market support mechanisms.

Subsidies applied to both on- and off-grid markets. The major market until the late 1990s was the off-grid one. Figure 6.2 shows the relative proportions of on- and off-grid cumulative installation through the 1990s. It can be seen that it was not until 1996 that

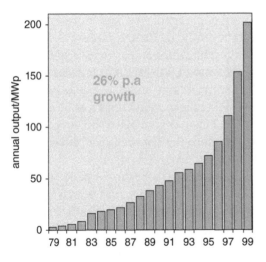

Figure 6.1   Growth of annual installations in the world photovoltaics market 1979–1999, MWp p.a. (*Source:* Courtesy BP Archive.)

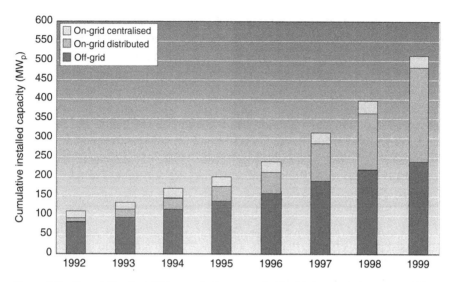

Figure 6.2 Cumulative installed capacity for on- and off-grid photovoltaics systems [8] (*Source:* IEA-PVPS-Trends 2000. Courtesy IEA PVPS.)

on-grid annual markets overtook the off-grid sector [8]. Most of the off-grid markets were in developing countries where capital subsidies were not relevant in stimulating the market. Spain and Norway did have a large number of unelectrified holiday homes, but this was not seen as a suitable market for subsidies.

As can be seen from Figure 6.2, the on-grid market was relatively small in 1992, but the sector began to grow markedly from that point on as the result of capital subsidies. The German 1000 Roof Programme began in 1990, as described in Chapter 5. The level of subsidy was high at 70% of initial cost, indicating how strong the push had to be to get the market moving [9]. This programme was so successful that it was oversubscribed and 2200 systems were installed. Similarly, Japan instituted a market stimulus programme in the early 1990s, with subsidies in the range of 50–67% of initial capital cost available for grid-connected systems [10]. In comparison, subsidies in the United States were more modest, with 35% tax exemptions for grid-connected systems. Reflecting this, Japan was the largest global market up to 2003. In 1999, its market-stimulation budget was $149 million, in comparison to $46 million for Germany and $66 million for the United States.

Capital subsidies were useful in kickstarting the market but had a number of disadvantages. It was apparent that the level required to take the market to the hundreds of MWp point would be very high. This would be challenging for government budgets, and the timescale required for the impact to become apparent would extend beyond the sitting term of any elected government anywhere in the developed world. In addition, subsidies were seen as favouring the more affluent members of society, who could afford to invest in new, high-cost technology [7]. During the 1980s, the urgency of deploying renewable energies on a large scale also diminished. The driving force behind the exploitation of renewable energy in the 1970s, as described in Chapter 2, had been the avoidance of future oil shortages and severe price rises. Through the 1980s, the oil price fell from a peak of $35/barrel in 1980 to below $10/barrel in 1986 (unadjusted for inflation) due to a glut in oil supply resulting from energy-saving measures and the discovery of new resources [11]. Therefore, the urgency of accessing new energy supplies decreased significantly. Only as the concerns shifted to climate change abatement in the 1990s did a serious commitment to exploit the potential of renewable energies resume.

6.3   Feed-in-Tariffs

A further disadvantage of capital subsidies was the fact that there was no requirement for the subsidised photovoltaic system to operate well and to continue to operate. No value was put on the electricity produced. Early FIT schemes attempted to fix this, but they were not particularly effective as incentives. For example, from 1990 in Germany, system owners could feed excess electricity from a photovoltaic system into the grid and be paid between 75 and 90% of the retail electricity rate [12]. Other countries would give 100% of the retail rate. This did not constitute a significant incentive. The market dynamic was totally changed with the introduction of the 'Aachen' model in 1994. As described in Chapter 5, when the 1000 Roof Programme came to an end, there were no further incentives in place to replace it, and the photovoltaic industry in Germany suffered a severe decline. The Aachen model proposed that owners of photovoltaic systems should be able to sell their electricity to the grid at a rate which enabled them to recover the cost of installing their system [13]. This would be funded by all other electricity ratepayers paying a small amount more. The introduction of this scheme was made possible because many utilities in Germany were controlled by the regional governments, which were responsive to their electorates. In Aachen, the sum paid to photovoltaic system owners was 2DM/kWhr ($1.34/kWhr). Such a high rate could be offered because the 1990 law specified a minimum price for photovoltaic electricity but not a maximum one. The additional charge to the ratepayer was capped at 1%. Other regions subsequently put the cap as high as 3%. The cap in Aachen equated to the installation of 1.5 MWp. Although this seems small in modern terms, at the time the total world market was only 50 MWp (Figure 6.1), so that this single relatively small city could make up 3% of global installation.

Compared to capital subsidy, six advantages can be identified with the FIT model [13]:

1) *Stable funding source.* The finance was provided by all electricity users, which for domestic consumers in Aachen amounted to no more than $0.4 per month on their bill. There was no charge against government budgets, and the rate could be guaranteed for 20 years, making for a stable investment scenario.
2) *Stable market for manufacturers.* A stable and growing market provided the security for manufacturers and installers to invest in the capacity to facilitate a

long-term business. This increased capacity and investment in new technology led prices to fall.

3) *Demand based on public choice.* The FIT scheme freed the photovoltaics market from government budget restrictions so that it could grow at a rate supported by the general public: a democratisation of photovoltaics roll-out.

4) *Competitive pricing for photovoltaic system customers.* As the purchase decision rested with the owner, the FIT provided a competitive market which encouraged installers to offer the lowest prices.

5) *Improved system efficiency.* Capital subsidies has focussed on the purchase price, but the FIT scheme promoted efficient system operation and longevity. This pointed the way to improved system design, better installation and practice, and more efficient inverters. It also directed module manufacturers to optimise kWhr/kWp output, and this has become a significant metric in determining module quality and the viability of new cell technologies.

6) *Widening the investment community.* The potential for long-term returns from FIT made photovoltaic installations a suitable opportunity for utilities companies and others investors, greatly increasing the finance available for deployment.

FIT became a hugely successful vehicle for developing the photovoltaics market in Germany, and subsequently in many other countries. The Aachen model filled the gap when the 1000 Roof Programme ended. It provided a mechanism for solar advocates to initiate incentive programmes in their own region. Within two years of the Aachen roll-out, a further 30 cities and regions had introduced or were in the process of introducing their own FIT provisions, including Berlin, Frankfurt, Munich, and Hamburg. As shown in Figure 6.3, by 1999, 46 cities had FIT programmes with rates greater than 0.8DM/kWhr [14]. Although a total of 14 MWp was installed under these local initiatives, individual caps were soon reached and a national programme was needed.

The success of the Aachen model showed the strong public support across Germany for FIT. Two initiatives overlapped. To follow up on the 1000 Roof Programme, a 100 000 Roof Programme was approved by the German parliament in 1999, but this was an interest-free loan scheme and it was apparent that a national FIT was also needed at a level capable of stimulating the market [15]. The Renewable Energy Law (EEG) was enacted in 2000, introducing a FIT of 0.99DM/kWhr ($0.52/kWhr). A range of photovoltaic systems qualified for the EEG, including both rooftop-mounted and ground-mounted installations. Rooftop systems below 30 kWp received the highest FIT, while systems up to 1 MWp were granted lesser ones. A new market was created in applying the FIT to ground-mounted agricultural systems, typically in the range of hundreds of kWp. The FIT would be reduced with time as photovoltaic system costs fell. The introduction of the national scheme was highly successful in stimulating the market at no additional cost to the government budget. As seen in Figure 5.8, the German market grew explosively with the introduction of the FIT, with growth in some years exceeding 100%; a total of 24.7 GWp was installed by 2011. The most popular sector was the agricultural and commercial one, so that by 2018 58% of systems

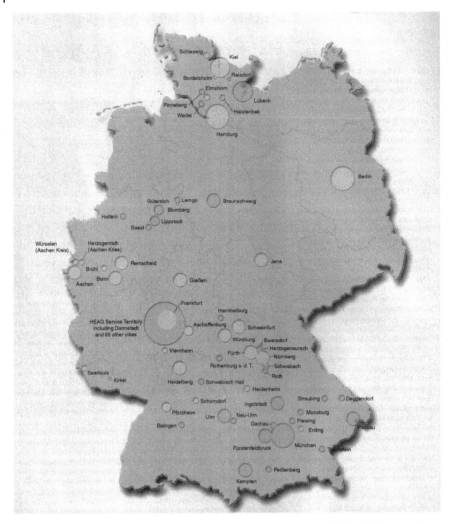

Figure 6.3 German cities offering significant FITs (circle area) in 1999 [14] (*Source:* A.Kreutzmann: Photon International 9 (1999) 25–26. Courtesy Photon International.)

were in the range 40–750 kWp [16]. The FIT had been systematically reduced as photovoltaic system processes fell and installations increased. In its most recent form, the FIT ranged from €0.11/kWhr for installations <10 kWp to €0.085/kWhr for ones ≤750 kWp, compared to a retail electricity price of around €0.3/kWhr [16]. Self-consumption was subsidised up to 2012 as a premium payment above the retail electricity price, but this was dropped when the FIT fell below that price. Instead, subsidies were given for battery storage systems, to enable self-consumption [17]. However, the FIT in Germany had served its purpose in driving down costs and providing for a significant part of the nation's electricity-generating capacity.

Table 6.1 Global growth in incentives for renewable energy [19]

	Start 2004	2013	2014
Countries with policy targets	48	144	164
States/provinces/countries with FITs	34	106	108
States/provinces/countries with renewable portfolios/quotas	11	99	99
Countries with tendering competitive bidding	n.a.	63	64

Source: KPMG: Report "Taxes and Incentives for Renewable Energy" 2015, KPMG International. © 2015, KPMG International Cooperative.

The success of the FIT scheme in Germany resulted in its replication in other countries. Italy had a FIT in place from 2005 to 2012, which boosted its market considerably. France maintains a FIT for systems below 100 kWp and has put special emphasis on BIPV with an additional subsidy. Spain introduced a FIT, but it was withdrawn in 2008 when the market boomed too rapidly. The United Kingdom introduced a FIT in 2010, and from a very small base 1 GWp was installed in a year. At this time, the European Commission concluded that 'well adapted FIT regimes are generally the most efficient and effective support schemes for promoting renewable electricity' [18]. FIT continues to be a major means for market support in many countries. Table 6.1 details the growth in incentives for renewable energy. It shows both the importance attached to stimulating the deployment of renewables and the preference for FIT as a means of promoting photovoltaics deployment. Currently, China is the world's largest photovoltaics market, and this has been achieved using the FIT model for both rooftop and large utility-scale systems [17]. Japan is also operating a FIT scheme.

The initial concept of the FIT in the Aachen model was very sound in enabling a rapid growth in the photovoltaics market, with all electricity users covering the cost. Challenges emerged, however, as the number of photovoltaic systems increased and prices fell, so that ultimately the levelised cost of electricity (LCOE) from photovoltaics was comparable to the retail electricity price. Managing the level of FIT with rapidly falling photovoltaics prices proved problematic [20], with surges in the market when the FIT had not adequately tracked falling costs. The connection of multiple small photovoltaic systems to the grid also caused oversupply of electricity, especially in the midday period, requiring intervention in remotely closing photovoltaic systems. Thus, various schemes for encouraging self-consumption were introduced. In addition to subsidising storage, as in Germany, limits were placed on the amount of photovoltaic electricity that could be fed into the grid. Many countries, such as Spain, allowed net-metering to encourage self-consumption in reducing electricity bills, although complications arose in the application of taxes and grid maintenance costs to self-consumed electricity. Nevertheless, the future for smaller systems of less than 100 kWp appears to be simply a reduction in electricity costs for the system owner [20], rather than the offering of any subsidies such as FIT. Larger systems require different regulatory and remuneration frameworks.

6.4 Power Purchase Agreements and Other Incentives for Large-Scale Systems

Chapter 5 detailed the arguments in favour of photovoltaics as a distributed electricity-generating source. To summarise: the demand for electricity is dispersed, as is the solar resource. The distributed on-grid market was stable at around 16–19 GWp p.a from 2011 to 2016 but surged to 36.8 GWp in 2017 as China expanded this sector [20]. Nevertheless, the centralised on-grid photovoltaics market grew much faster, with 61 GWp of the 98 GWp installed in 2017 being centralised grid deployments. The relative proportions of annual installations are shown in Figure 6.4. It can be seen that from 2013 onward, the majority of photovoltaic systems installed globally were centralised on-grid (i.e. utility-scale) systems. It is interesting to note the cumulative installations, as shown in Figure 6.5. By 2017, nearly 60% of the global photovoltaic capacity was in utility-scale systems. This distribution is also a consequence of the rapid growth of the markets in Asia as compared to the traditional markets in the United States and Europe, as shown in Figure 6.6. It can be seen that it is only in Europe that the distributed photovoltaic on-grid sector is bigger than the utility-scale deployment. This largely reflects needs in Asia, where there is a deficit in electricity production in the rapidly expanding economies of China and India, and in the United

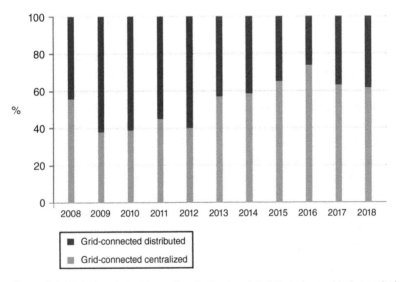

Figure 6.4 Relative proportions of centralised and distributed on-grid photovoltaic generation by year from 2007 [20] (*Source:* IEA PVPS trends 2019. Courtesy IEA PVPS.)

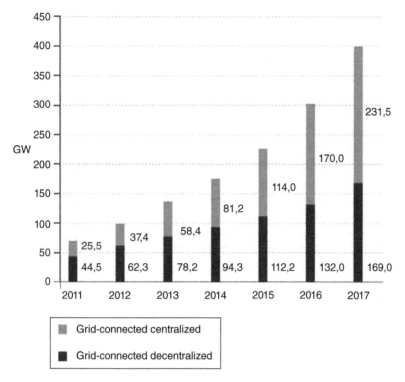

Figure 6.5 Cumulative installation of centralised and distributed on-grid photovoltaic generation by year from 2011 [21] (*Source:* IEA PVPS Trends 2018. Courtesy IEA PVPS.)

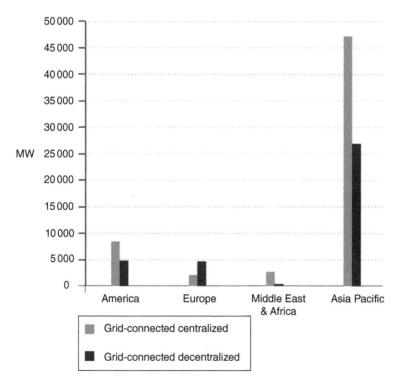

Figure 6.6 Installation of grid-connected centralised and decentralised photovoltaic systems by region in 2017 [21] (*Source:* IEA PVPS Trends 2018. Courtesy IEA PVPS.)

States and Japan, where there is a drive to reduce carbon dioxide emissions. These outcomes are the results of the various incentive programmes that have been implemented.

As photovoltaic system sizes have increased, this has given rise to two issues: the cost of FIT for large systems and the integration of the power generated by the plants. For example, the Chinese government dramatically cut subsidies in May 2018, in part because the national target of 105 GWp by 2020 had been achieved by 2018, although distributed photovoltaics continued to be supported. Governments have sought to control the growth of large plants through the tendering of power purchase agreements (PPAs) [21]. In 2017, 12% of all photovoltaic installations were achieved by tender, and this percentage is growing. Meanwhile, in 2018, 32% of all utility-scale plants were funded through a tendering process [20].

In a tender, government agencies offer a certain electricity capacity for a fixed period and photovoltaic system developers offer installations at some fraction of this at a price for the electricity produced. PPA is a variant of this process whereby the photovoltaic system developer owns, operates, and maintains a photovoltaic system and a host customer agrees to site that system on their property and purchase the system's electric output from the provider for a predetermined period. Frequently, in both tenders and PPAs, the contract is sold to an investor, who takes a guaranteed long-term return form the photovoltaic generation while the developer continues to operate and maintain the solar field. These arrangements can be quite complex, as illustrated in Figure 6.7, which shows the structure of a typical solar PPA in the United States [22].

Such an auction-based method of funding large photovoltaic systems has been particularly effective in reducing the cost of electricity from photovoltaic plants. A particular example is Chile, where, in the northern region, there is good access to the national grid, low-cost land, and a good solar resource, giving a yield of 3000 kWhr/kWp. In 2016, this led to a winning tender for 720 GWhr of production at $0.0291/kWhr, a world record at the time, while in 2017 an even lower bid of $0.0254/kWHr was awarded for 900 GWHr [20]. The El Romero Solar photovoltaic field is shown in Figure 6.8. This plant has a nominal peak capacity of 246 MWp and has been operational since 2016, when it was the largest photovoltaic installation in Latin America [23]. The low price for power was achieved in a good solar regime.

The Middle East has also posted some very low tenders. Table 6.2 shows the range of the most competitive tenders up to the first quarter of 2018 [21], although tender prices continue to fall. In mid-2019, Brazil awarded 203.7 MWp to photovoltaics, with the lowest bid at $0.0169/kWh [24]; this was matched by similar tenders in Dubai. Subsequently, a record-breaking bid of $0.01646/kWhr in Portugal gave way to a lowest winning bid at the end of 2019 of $0.01567 for an 800 MWp plant in Kharsaah, Qatar [25]. Of course, these low bids are for areas with very high solar irradiances and low population densities, so that land prices are not an issue. The Atacama Desert in Chile, for example, is one of the driest places in the world. More remarkably, in Germany, tenders for photovoltaic generation below $0.0355/kWhr [26] are well

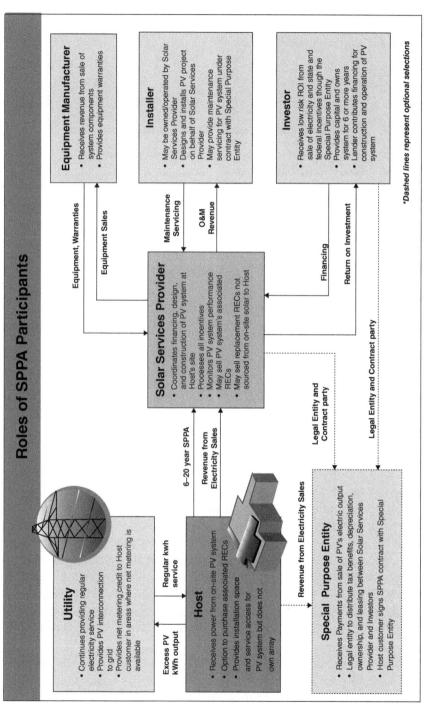

Figure 6.7 Structure of a typical solar PPA, as described by the US Environmental Protection Agency [21] *Source:* www.epa.gov/greenpower/solar-power-purchase-agreements accessed 22nd March 2019

Figure 6.8 El Romero Solar Field, Atacama Desert, Chile (*Source:* Acciona Energia)

Table 6.2 Most competitive tenders for the supply of photovoltaic electricity globally up to Q1 2018 [20]

Region	Country/state	US cents/kWhr
Latin America	Mexico	2.057
Latin America	Chile	2.148
Latin America	Brazil	3.558
East Asia	India	4.1
Latin America	Argentina	4.04
India	India	4.6
Middle East	Armenia	4.6
Western Europe	Germany	4.87

Source: Modified from IEA PVPS trends 2019

established. This compares favourably with the wholesale electricity price in Germany, which in 2018 averaged $0.0505/kWhr [27].

Tenders and PPAs are quite blunt instruments, but some shaping of the outcomes can be undertaken by governments in order to obtain specific outcomes. The competitive nature of the sector has been successful in driving down the price of electricity from photovoltaic generation. To give more focus to strategic developments, China introduced its 'Front Runner' programme in 2015 to stimulate the manufacture of high-efficiency modules. In 2017, 10 GWp of installation was reserved for such modules, with a minimum multicrystalline silicon module efficiency of 17% and a minimum monocrystalline one of 18%. Small segments of 50 MWp were set aside for the specific technologies of PERC, SHJ, and HIT [28]

In France, the FIT is structured to reflect the difference in solar regimes from the north to the south of the country, as well as targeting BIPV as a sector for stimulation [20]. In the United States, a rather different support mechanism is used. Regulation of the electricity industry is at the state level. About half the market is in distributed

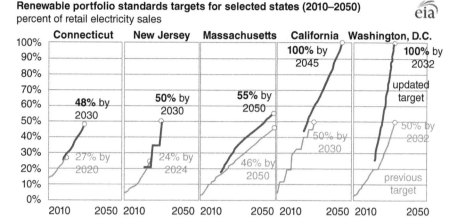

Figure 6.9 RPS targets for four leading US states plus Washington, DC [29] *Source:* www.eia.gov/todayinenergy/detail.php?id=38492 accessed 1st April 2020

grid-connected projects, where net metering and tax credits are the main support mechanisms. Utility-scale projects are underpinned by the requirement to meet Renewable Portfolio Standards (RPS). These instruments are applied at the state level and require that the utilities companies supply a certain proportion of electricity by renewable means, with photovoltaics one of several accepted technologies. PPAs are then put in place to meet these requirement. RPS are in place in 29 states plus Washington, DC; Figure 6.9 shows their evolution in five leading examples [29]. It can be seen that the targets are ambitious, particularly for Washington, DC and California. In the latter, the objective is 50% renewables by 2030, rising to 100% by 2045; in the former, an initial target of 50% renewables by 2032 has been adjusted to 100%.

The effect of the various subsidy programmes has been to create a marketplace where utility-scale plants are now the dominant and growing market sector. The trend was apparent as early as 2004, when this author calculated a 300% increase in large power plants >10 MWp between 2003 and 2005. Although subsidies for such plants are now being decreased, it seems the greatest market for photovoltaic installations in the future lies in utility-scale applications.

6.5 Summary

This chapter has mapped how subsidies facilitated the development of photovoltaics from a small high-cost industry with little global impact to a large multinational operation that has reduced costs in many countries to be fully comparable with other forms of electricity generation, including nuclear and fossil fuel. The subsidy programmes felt their way and developed much as photovoltaics manufacturing itself did. Initially, capital subsidies helped, but the development of FIT was crucial to the expansion of the market which gave investors confidence to create the GWp scale of manufacture, bringing costs down to such a level that utility-scale photovoltaic installation could be achieved and the photovoltaic technology could become truly competitive as a global electricity production technology. The subsidy programmes have thus been highly effective.

References

1 W. Palz: 'Solar Power for the World' pub: Pan Stanford (2014) 57.
2 M.G. Coleman and L.A. Green: Proceedings of the 15th IEEE PVSC (1981) 713–717.
3 T.M. Bruton et al.: Proceedings of the 14th EUPVSEC, Barcelona (1997) 672–677.
4 C. del Canizo, G. del Cosa, and W.C. Sinke: Progress in Photovoltaics 17 (2009) 199–209.
5 Photon International 11 (2018) 60.
6 W. Palz: 'Solar Power for the World' pub: Pan Stanford (2014) 56.
7 R.J. Gilbert: 'Regulatory Choices: A Perspective on Developments in Energy Policy' pub: University of California Press (1991) 293–296.
8 IEA-PVPS-Trends 2000.
9 W. Sandtner in 'Solar Power for the World' ed. W. Palz pub: Pan Stanford (2014) 335–339.
10 IEA PVPS Report 1995 53.
11 https://en.wikipedia.org/wiki/1980s_oil_glut accessed 7 August 2020.
12 H. Scheer in 'Solar Power for the World' ed. W. Palz pub: Pan Stanford (2014) 293.
13 A. Kreutzmann, W. von Wolf, and P. Welter: IFAC Proceedings 28 (1995) 151–153.
14 A. Kreutzmann: Photon International 9 (1999) 25–26.
15 H. Scheer in 'Solar Power for the World' ed. W. Palz pub: Pan Stanford (2014) 297.
16 Photon International 12 (2019) 26–31.
17 IEA PVPS Trends 2017.
18 European Commission Staff Working Document January 2008.
19 KPMG: 'Taxes and Incentives for Renewable Energy' (2015).
20 IEA PVPS Trends 2019.
21 IEA PVPS Trends 2018.
22 https://www.epa.gov/greenpower/solar-power-purchase-agreements accessed 7 August 2020.
23 Photon International 1 (2017) 28.
24 Taiyang News 8 July 2019.
25 Taiyang News 25 January 2020.
26 Taiyang News 24 February 2020.
27 Clean Energy Wire 13 November 2018.
28 https://www.solarquotes.com.au/blog/chinas-top-runner-program/ accessed 7 August 2020.
29 https://www.eia.gov/todayinenergy/detail.php?id=38492 accessed 7 August 2020.

7

Difficulties of Alternative Technologies to Silicon

7.1 Introduction

Chapter 4 laid out the advantages of silicon as a solar cell material and recorded how it has been the dominant semiconductor material for photovoltaic applications for the past 40 years and is expected to continue to be so for some time. This is somewhat surprising given that the one inherent disadvantage of silicon is that it is an indirect bandgap semiconductor, and therefore relatively thick wafers of at least $100\,\mu m$ are required to make good-efficiency solar cells. Therefore, other technologies have been explored which utilise a combination of less semiconductor material with direct bandgap materials, easier fabrication, and simpler module construction, resulting in cost savings. The three principal approaches are as follows:

1) Silicon sheet processes, in which silicon is formed directly into a wafer or sheet without going through a wafering process.
2) Thin films (typically $1\,\mu m$ thickness) of a direct bandgap semiconductor deposited on to glass, steel, or plastic substrate in large areas. Candidate materials include copper sulphide, amorphous and microcrystalline silicon, cadmium telluride, copper indium gallium diselenide, dye-sensitised solar cells, organic semiconductors (OPVs), and perovskites.
3) Concentrator technologies, in which lenses or mirrors focus sunlight by a factor of $2\text{--}1000\times$ on to a small-area solar cell. Initially, silicon was used for this application, but it has been displaced by stacked layers of III–V material forming a multijunction solar cell, frequently on a germanium substrate.

7.2 Sheet Silicon Processes

As early as 1973, it was recognised that Czochralski (CZ) silicon had limitations in terms of cost reduction and therefore research into alternative sheet silicon technologies was necessary [1]. The driving force behind sheet silicon processes was the

Photovoltaics from Milliwatts to Gigawatts: Understanding Market and Technology Drivers toward Terawatts, First Edition. Tim Bruton.
© 2021 John Wiley & Sons Ltd. Published 2021 by John Wiley & Sons Ltd.

elimination of losses in casting and wafering, which could be up to 50% in multicrystalline silicon and 60% in monocrystalline crystal growth. Three distinct approaches have been followed:

1) The direct crystallisation of near single-crystal material direct from molten silicon.
2) Direct casting of liquid silicon into a mould or on to a low-cost substrate.
3) Kerfless wafering of silicon.

The cost reduction potential is highest for option 1 and lowest for option 3, which requires silicon material that has gone through a crystallisation process.

7.2.1 Direct Crystallisation of Silicon Sheet

One of the challenges of growing a sheet of silicon directly from the melt is in stabilising the edge of the growing film. In the dendritic web technique, the dendritic crystals defining the liquid film provide the stability. In the edge defined foil growth (EFG) process, hollow shapes (initially nonagons, later octagons) are grown. Finally, in the string ribbon process, two metal ribbons are used to draw the film of liquid silicon from the melt. These approaches are discussed in turn.

7.2.1.1 Westinghouse Dendritic Web

The earliest process utilised was the dendritic web, pioneered at Westinghouse [2]. A seed crystal is introduced into a supercooled liquid silicon and spreads laterally to form a button. As the button is withdrawn from the melt, two dendritic crystals grow rapidly down into it; as they are pulled through the melt, a film of liquid silicon is drawn up, which solidifies rapidly. The silicon thus formed is nearly single-crystal, although it is heavily twinned. The process is illustrated in Figure 7.1.

The rapid growth means segregation of impurities does not take place and the grown material has a much lower lifetime than CZ material, plus a high oxygen content. This limits the efficiency that can be obtained. In the mid-1970s, it was just possible to achieve a 10% cell efficiency [3]. Development at Westinghouse continued until 1994, when Ebara Corp (Japan) acquired the technology and continued to refine it as Ebara Solar Inc. in Pittsburgh, PA. By 1997, a record 17.3% efficiency had been achieved on a $25\,cm^2$, $100\,\mu m$-thick, n type wafer [4]. However, a relatively slow growth rate and limited width of the ribbon made the approach less competitive than other ribbon technologies [5]. Ebara Solar Inc. ceased operations in 2002 [6].

7.2.1.2 Edge Defined Foil Growth

The EFG ribbon process was developed at Tyco Corporation [7] and later acquired as part of the formation of Mobil Solar. The process is illustrated in Figure 7.2. It works by immersing a slotted die made of carbon into a silicon melt. A film of liquid silicon is drawn up through the die, forming a liquid meniscus. A single-crystal seed is used to contact the meniscus and then slowly withdrawn. A film of liquid silicon is drawn up and solidifies. Film thicknesses between 100 and $400\,\mu m$ can be produced. In early developments, ribbons of 25 mm width and up to 20 m length could be produced, with

Figure 7.1 Growth of a silicon ribbon by the dendritic web process [5] (*Source:* W. Koch et al: Handbook of PV Science and Engineering J. Wiley eds: A Luque & S Hegedus, 239. © 2003, John Wiley & Sons. Courtesy Wiley.)

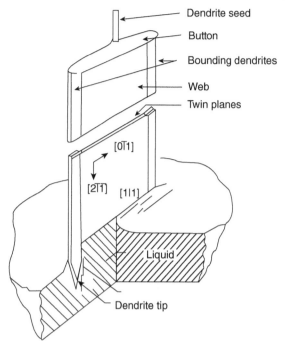

AM0 solar cell efficiencies for small-area cells approaching 10% [7]. The ribbon width later increased to 50 mm, and by 1981 multiple ribbons could be grown at 10 cm wide in one furnace [8]. Stabilising the edge of the ribbon was a problem, but it was solved by growing hollow forms. In 1983, nine-sided figures with 5 cm-wide faces were grown, and in 1988, this was expanded to an octogon with 100 mm-wide faces. A 12% cell efficincy was achieved in 1988 [8], but hydrogen passivation was essential to overcome the high levels of defects in the material. By 2002, 14% cell efficiencies on 100 cm² could be achieved [9], while with advanced processing efficiencies, up to 16.7% (4 cm²) could be obtained [10].

As the technolgy developed, industrial-scale production took place – together with several changes of ownership. From the original Tyco development, Mobil Tyco operated from 1974 until 1982, when the company became Mobil Solar. This then passed into the ownership of ASE (an RWE subsidiary) in 1993.

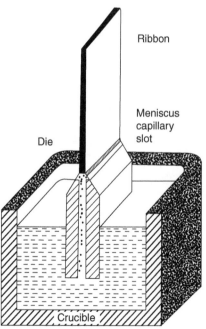

Figure 7.2 Schematic of EFG growth equipment [5] (*Source:* W. Koch et al: Handbook of PV Science and Engineering J. Wiley eds: A Luque & S Hegedus, 239. © 2003, John Wiley & Sons. Courtesy Wiley.)

ASE expanded production in the United States from 4 to 20 MWp p.a. by 2001 and started a further 20 MWp capacity at the Alzenau, Germany site [9]. In 2002, ASE (now RWE Solar) formed a joint venture with Schott Glas to become RWE Schott Solar [11], and in 2005 Schott became the sole owner. The EFG technology faced several challenges due to limitations in the EFG material quality, which required complex processing, and surface roughness, which required a nonstandard metallisation technique. The limiting factor, however, was the rate of production of silicon area. Pulling rates of the ribbon remained low – below 2 cm/min, comparable to CZ growth rates. Calculations showed that cyliders of 1 m diameter were needed to give truly competitive production rates [12]. Although ones up to 0.5 m were grow, solar cells could not be successfully fabricated. The same production line at Alzenau that gave 14% efficiency on EFG wafers yielded 15% efficiency on conventional multicrystalline ones. Schott Solar announced the closure of the plant in 2012 [13].

7.2.1.3 String Ribbon Technology

By the early 1980s, it was recognised that both the dendritic web and the EFG approaches had important limitations. A new approach was thus developed jointly by Arthur D. Little and MIT, initially named the edge stabilised ribbon growth technique and later known as the string ribbon technique [14]. The process is illustrated in Figure 7.3. Metal strings are used to stabilise the foil edge. These are drawn through the silicon melt, pulling up a film of silicon, which solidifies. In principle, several pairs of strings can be drawn through the same crucible. Early developments gave 11.0%-efficient solar cells on a 2.2 inch-wide ribbon (105 cm^2) [15].

The next major step toward commercialisation was the formation of the Evergreen Solar Company in the United States in 1994. By 2002, the process had been improved such that ribbon growth was continuous, with sections 2 m in length being laser-cut and active cells 8 cm wide and 15 cm in length being fabricated [16]. The best production cells were 15.4% efficient, while the best laboratory cell could achieve 17.7% (4 cm^2) efficiency [17]. In 2006, Ever Q was formed in Germany as a joint venture with Q Cells, Evergreen, and the Renewable Energy Corporation. The aim was to establish a 300 MWp annual production. However, strong competition from China caused Evergreen to go into chapter 11 bankruptcy protection in 2011 [18], and Ever Q (by then named Sovello) ceased production in 2012 [19].

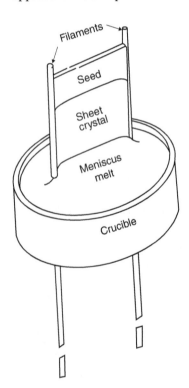

Figure 7.3 Growth of a silicon string ribbon [5] (*Source:* W. Koch et al: Handbook of PV Science and Engineering J. Wiley eds: A Luque & S Hegedus, 239. © 2003, John Wiley & Sons. Courtesy Wiley.)

7.2.2 Cast Silicon Sheet

Approaches toward growing good-quality near single-crystal ribbons of silicon did not achieve commercial success due to the relatively low linear growth rates of the ribbons and the difficulty in obtaining ribbons of high enough quality to give high-efficiency solar cells. This provided an opportunity for other technologies for the direct casting of multicrystalline silicon wafers to come on the scene.

7.2.2.1 Hoxan Casting Process

One of the limitations in the ribbon growth techniques was that the rate of sheet crystallisation was determined by the linear growth rate of the ribbon. Hoxan developed a spinning method in which liquid silicon could be forced into silica moulds 100 mm square, allowing sheets of between 100 and 500 μm to be produced [20], which decoupled the crystallisation rate from the supply of liquid silicon. The mould was changed to graphite for better reuse, and a silicon nitride coating was applied to prevent liquid silicon from adhering to it. Four wafers could be cast simultaneously in 15 seconds and 10%-efficient (4 cm^2) solar cells were produced [21]. Disassembling the mould limited production, and the technique was modified to draw a multicrystalline sheet from the mould [22]. Initially, 300 μm-thick sheets of up to 100 mm width could be produced at pulling rates of 200–400 mm/min, but the process was subsequently improved to allow 200 mm-wide sheets to be fabricated. A 100 cm^2 solar cell cut from the larger sheet gave efficiencies of 11.7% [23], comparable with other multicrystalline silicon solar cell technologies in 1992 [24]. No further progress appears to have been made.

7.2.2.2 Ribbon Growth on Substrate

Next came the development of the ribbon growth on substrate (RGS) process, in which a substrate is drawn underneath a rectangular reservoir and coated with a layer of liquid silicon, which then crystallises [25]. The process is shown in Figure 7.4. The separation between linear production rate, in which the substrate moves horizontally, and the crystal growth, which is vertical, is evident. The development was first started at the Bayer Company and then transferred to the Energy Research Centre of the Netherlands.

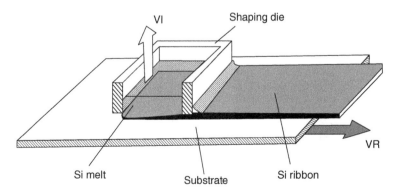

Figure 7.4 Schematic of the RGS process [5] (*Source:* W. Koch et al: Handbook of PV Science and Engineering J. Wiley eds: A Luque & S Hegedus, 239. © 2003, John Wiley & Sons. Courtesy Wiley.)

By 2006, cells of 13% efficiency were being produced on 5×5 cm RGS wafers [26]. This was sufficiently promising for a scale-up of the technology to take place, and a continuous 50 MWp p.a. pilot facility was established at RGS BV in the Netherlands [27]. The pilot machine produced 156×156 directly cast wafers with a throughput of 10 000 cm²/min. This machine gave solar cell efficiencies of 11.7%, while the best laboratory cells were 14.4%; this was relatively low compared to the then state of the art for conventional multicrystalline wafers. Recent research has been directed toward casting on to seed wafers to form quasi-monocrystalline ribbons with efficiency up to 16% [28]. While RGS Developments still exists and research on the ribbon material continues, there are currently no commercial sales of this material [29].

7.2.2.3 Direct Wafer

The most remarkable progress in direct casting of wafers has been made by the 1366 Technologies Company in the USA. This company was formed in 2008 and received funding from the US ARPA-E programme for its proof of concept [30]. A proprietary process based on casting directly on to a porous sheet in contact with a silicon melt enabled highly uniform stress-free silicon sheets to be cast and easily released [31]. The technology was developed by Prof. E. Sachs at MIT. By 2013, a fast, scalable process had been developed in order to produce 156 mm square wafers with minority carrier lifetimes over 100 μs, giving solar cell efficiencies over 17% – comparable with conventionally cast, wire-sawn wafers [32]. The progress of the technology is summarised in Figure 7.5. The company has progressively scaled up. In 2016, a contract was signed to supply 700 MWp equivalent of wafers to Hanwha Q Cells and additional capital funding was received from Wacker Chemie [33]. In 2017, a record 20.3%-efficient solar cell was produced at Hanwha Q Cells. The expectation for the direct wafer is that wafer cost will be halved and embedded energy reduced by 60%. This is to date the most successful kerf-free silicon wafer development.

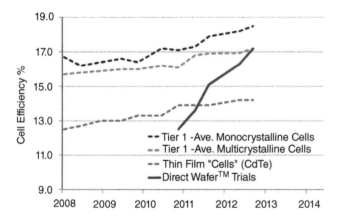

Figure 7.5 Progress in cell efficiency of direct wafer solar cells [32] (*Source:* E. Sachs et al: Proceedings - EU PVSEC 2013, 28th European Photovoltaic Solar Energy Conference and Exhibition, 907-910. © 2013, WIP GmbH & Co Planungs-KG. Courtesy EUPVSEC.)

7.2.2.4 Lift-Off Wafer Technology

In the mid-1990s, it was recognised that the electrochemical etching of silicon to produce a porous silicon layer could be used to fabricate thin silicon foils in which solar cells could be made [34,35]. An electrochemical process is used to progressively etch a monocrystalline silicon substrate in order to produce a deep, high-porosity layer and then a less porous layer at the wafer surface. Heating the wafer causes the surface to reorganise and form a dense surface layer, maintaining the single-crystal orientation, while a highly porous layer forms typically 1–2 µm below the surface. This has been termed a 'quasi-monocrystalline silicon' (QMS). The annealing process is carried out at around 800–1000 °C and can be achieved in a chemical vapour deposition (CVD) reactor prior to epitaxial deposition, which is required to give a sufficiently thick silicon layer for good solar cell efficiency [36]. A highly doped p type film can be deposited epitaxially on to the QMS in order to form BSF, followed by a thicker lightly doped absorber region. Surface diffusion and conventional metallisation may be applied to complete the front side. The cell is then bonded to a superstrate glass and the base wafer is removed to allow reprocessing for further QMS fabrication. The rear surface is metallised to complete the solar cell. Early results gave a 13.6% (4 cm²)-efficient solar cell with a 23 µm epitaxial layer [36]. Over time, the process was improved to give 19% cell efficiency in a 43 µm-thick cell [37]. Further development has led to an inline process development with surface reorganisation and high-deposition-rate epitaxy at 1150 °C [38]. 'Epi-Wafers' of between 40 and 150 µm were successfully fabricated in both p and n type with minority carrier lifetimes up to 1.5 ms for the n type material, which is high enough that solar cells with efficiencies in excess of 20% may be anticipated. This degree of progress has stimulated the formation of a new company, NexWafe GmbH, with a 5 MWp p.a. pilot line for the production of n type 'Epi-Wafers', with the first commercial production in 2018 and large-scale production set to begin in 2021 [39]. The advantage of this process is that it produces a wafer which is a direct replacement for conventionally wire-sawn wafers without requiring modifications to existing production lines.

7.3 Thin-Film Solar Cell Technologies

Given that the energy density of solar irradiance is normally at best 1 kW/m², large areas of solar cell are needed in order to generate significant power. It appears logical that, to generate large areas of solar cell, some kind of sheet process (ideally, inline roll-to-roll processing) should be used. For most of the past 40 years, it was expected that a new technology would emerge to replace the existing silicon wafer-based one, which relies on casting large blocks of silicon, sawing them into bars, slicing those bars into wafers, processing the wafers as solar cells, and laboriously interconnecting and laminating the wafers into modules in order to provide a large collecting aperture. The silicon technology is further handicapped by silicon's being an indirect bandgap semiconductor, such that solar cells must be made thicker than 100 μm to give a good efficiency. In contrast, a direct bandgap semiconductor would absorb the incident solar irradiation in less than 1 μm thickness of active solar cell.

A wide range of technologies are available for the deposition of semiconductor thin films, including printing, spraying, and vapour phase deposition in either vacuum or atmospheric pressure. Such processes are intrinsically capable of scale-up to large-area, high-throughput coating technologies. The transformation from solar cell to module can be done monolithically by scribing the semiconductor and applying the metallisation inline again. In principle, vacuum lamination should not be needed and environmental protection could be achieved by simply spraying a suitable polymer on to the rear of the solar cell structure, assuming a glass front surface.

A further benefit of the thin-film approach is that the energy embedded in producing the solar module is much less than in the case of silicon, and energy payback time is projected to be significantly less than one year [40]. By 1982, four thin-film technologies had achieved laboratory efficiencies of over 10% [41]. At that time, production silicon solar cells were around 12% efficient. It seemed that one or more thin-film technologies would thus shortly displace silicon. Indeed, many major corporations at the time had active thin-film research programmes, including BP, Shell, ARCO, Matsushita, Sanyo, Sharp, and Siemens. Nevertheless, nearly 40 years later, the reality is that only 5% of the world market is in thin-film technologies. This section reviews the development of the leading thin-film technologies and the reasons for their limited market penetration.

7.3.1 Copper Sulphide

Copper sulphide solar cells were introduced as the first real alternative to silicon, as described in Chapter 2. The initial cells were reported to be operational as early as 1954 [42] and were made by depositing a layer of cadmium sulphide and then converting a very thin layer of this by an ion exchange reaction to copper sulphide. The work was sponsored by the US Air Force Aero Propulsion Laboratory and subsequently by NASA, with the aim of producing low-cost, low-weight-per-kWp solar cells for space in competition with the dominant wafer silicon technology. The cost target was $10/Wp, as compared to the approximate $100/Wp for silicon [43].

The initial work in 1954 used single-crystal cadmium sulphide (CdS), but it was soon recognised that polycrystalline CdS, either sprayed or evaporated, could produce small-area cells of around 5% efficiency, and by 1965 the best cell of 50 cm^2 gave 8% efficiency under tungsten lamp illumination [44]. Despite this progress, cells made by the Clevite Corporation experienced problems with stability, losing up to 40% of output in one month [44]. By 1973, it was clear that these cells were unsuitable for space applications [45]. Nevertheless, research continued for the terrestrial market, with Böer's group at the University of Delaware taking a lead.

Böer had started studying CdS at the Humboldt University in Berlin, but joined the nascent group at the University of Delaware in 1968. By 1971, they were able to make an 8% efficient solar cell [46]. This was sufficient for the Shell Oil Company to invest $3 million in a pilot start-up as Solar Energy Systems Inc. While research continued, and eventually a 10%-efficient cell was produced in the laboratory [47], there were ongoing concerns over the stability of the Cu_2S/CdS solar cell, with three separate mechanisms for its degradation in output identified [48]. It was suggested that the lattice mismatch between the Cu_2S and CdS limited the cell efficiency to 10%, which made the technology uncompetitive with the other thin-film technologies emerging in the early 1980s [49].

7.3.2 Amorphous Silicon

Amorphous silicon was hailed as the first commercial thin-film photovoltaic technology [50]. The timing for this approach was good, as the demand for terrestrial photovoltaics was growing rapidly. The beginnings of the amorphous silicon technology in the 1970s were described in Chapter 2, but an explosion of activity occurred in the early 1980s. The first successful amorphous silicon solar cell was reported in 1976 [51]. Amorphous silicon is a form of silicon deposited at low temperatures such that the classic diamond structure is not realised. There is some short-range order, but no long-range, and there are many dangling bonds, as shown in Figure 7.6. In solar cell applications, amorphous silicon is generally deposited from silane by plasma-enhanced chemical vapour deposition (PECVD) at temperatures below 400 °C. The dangling bonds are passivated with hydrogen so that the material can exhibit photoconductivity. Amorphous silicon has a number of advantages as a solar cell material in that it is a

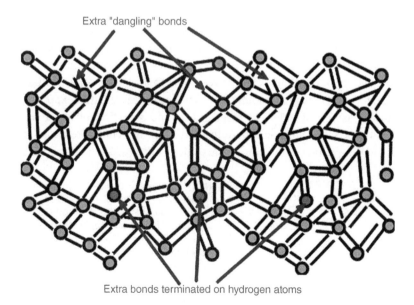

Figure 7.6 Amorphous silicon structure (*Source:* Courtesy pveducation.org)

direct bandgap semiconductor with a bandgap of 1.7 eV and it can be doped with germanium, further reducing the bandgap to 1.4 eV. At higher germanium concentrations, the material is too disordered for effective electron and hole transport. There are no lattice mismatch issues, and amorphous silicon can be grown on a wide range of substrates, including glass, ceramic, stainless steel, and plastic [50]. There are also no constraints on growing multiple layers, so that tandem cells with different bandgaps can be easily fabricated.

After the first demonstration of meaningful solar cell efficiencies, there was a rapid growth in research activity, with a range of possible amorphous silicon solar cell architectures embracing the dominant p-i-n structure, metal–thin-film insulator–semiconductor (MIS), and platinum/amorphous silicon Schottky diodes [52]. Indeed, in 1980, the best reported cell was an MIS device with a hydrogen and fluorine passivating the dangling bonds [53]; its efficiency was 6.3% under an AM1.0 spectrum. A typical p-i-n single-junction solar cell is shown in Figure 7.7.

While a range of cell structures were investigated, a decisive development was made at the University of Osaka in 1981 when the window p layer was changed to amorphous silicon carbide with a much wider bandgap, admitting more light into the i layer absorber and thereby increasing the efficiency to as high as 7.14% [54]. By 1982, a 10% efficiency had been reported for the first time [55] on a 1.0 cm^2 cell with a silicon carbide window. Given the rapid progress from 2% efficiency in 1976 to 10% in 1982, amorphous silicon was seen as a likely winner as the long-term technology choice for photovoltaics. A worldwide research effort sprang up, with laboratories in the United States, Europe, and Japan. Where only one group had reported a 10% solar cell efficiency in 1982, 25 laboratories had done so by 1989 [50].

Despite this rapid progress, there were problems with the technology. The p-i-n structure was essential to creating an electric field across the absorbing i layer in order to separate the photogenerated electrons and holes and transport them to the collecting electrodes. However, this field was weak in the centre of the i layer, limiting its thickness to 0.5 μm, which is less than is required to absorb all of the solar irradiation [52]. An additional problem was that solar cell efficiency decreased after exposure to light; this is known as the Staebler–Wronski effect, and was first observed in 1977 [56]. When electrons and holes recombine in the i layer, the energy released is sufficient to create dangling bonds in the amorphous silicon structure, which then act

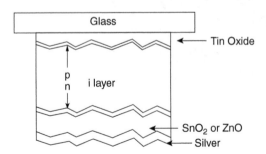

Figure 7.7 Schematic of a single-junction p-i-n amorphous silicon solar cell [50] (*Source:* R. Crandall and W. Luft: Progress in Photovoltaics Science and Applications 3, 315-331. © 1995, John Wiley & Sons. Courtesy Wiley.)

as traps for the photocarriers, lowering both short-circuit current and fill factor. In early devices, a reduction of efficiency as great as 50% could be observed [50]. With time, better devices were developed, and degradation of around 20% was considered typical. Record efficiencies are now reported for stabilised solar cells, with 10.1% for a single junction and 14.0% for a tandem cell after 1000 hours of exposure at 1 sun and 50 °C [57]. The i layer thickness plays a key role: generally, the thinner the layer, the more stable the solar cell [50].

The two constraints just mentioned shaped the development of the amorphous silicon cell. Thinner i layers were required for better current collection and improved stability, but they came at the cost of lower light absorption. Therefore, the structure of the front transparent conductive oxide (TCO) was important in order to increase light trapping and light absorption. Tandem cells were used to both increase the effective i layer thickness and permit different bandgaps to be used to broaden the portion of the solar spectrum that could be captured. Considerable research was directed toward improving light trapping. For cost reduction reasons, fluorine-doped tin oxide replaced indium tin oxide as the TCO of choice and much work went into developing a rough surface texture to maximise light coupling from the TCO into the amorphous silicon [58], although too rough an interface caused shunting problems in the cell as other constraints demanded thinner i layers [59]. With a limit reached on light trapping, more emphasis was put on the development of tandems with either two or three cells stacked. In the two-cell case, the top cell was amorphous silicon and the lower cell an amorphous silicon germanium alloy. The spectral response of an a-Si/a-Si/Ge tandem cell is shown in Figure 7.8. The upper-cell bandgap was 1.8 eV while the lower-cell

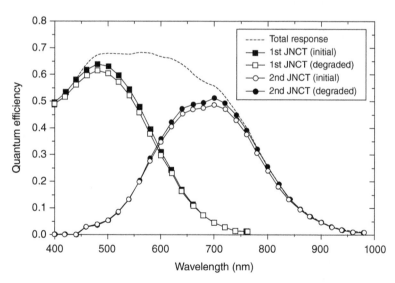

Figure 7.8 Spectral response of a typical stabilised a-Si/a-Si/Ge tandem solar cell [60] (*Source:* R.R. Arya and D.E. Carlson: Progress in Photovoltaics Science and Applications 10, 69-76. © 2002, John Wiley & Sons. Courtesy Wiley.)

bandgap was around 1.5 eV [60]. This is the lowest bandgap that can be achieved with germanium doping – at higher germanium levels, photoconductivity decreases rapidly. In the three-cell case, there were two stacked amorphous silicon cells with thin i layers and a bottom cell of amorphous silicon germanium. The best reported efficiency for a small-area triple-stacked all-amorphous cell was 13.7% ($0.25 \, cm^2$) [61].

In order to further increase efficiency and improve stability, research was directed toward developing hydrogenated microcrystalline silicon layers (μc-Si:H). The microcrystalline silicon consisted of silicon nanocrystals embedded in a passivating amorphous silicon matrix, with the nanocrystals forming a continuous conducting network. The μc-Si:H cell was stable against light degradation [62], provided the crystalline fraction was greater than 45%. The first fully microcrystalline cell, produced in 1996, gave an efficiency of 7.7% [63]. The combination of amorphous silicon with a bandgap of 1.8 eV and microcrystalline silicon with a bandgap of 1.1 eV (the 'micromorph' cell) is very near the maximum efficiency for a tandem solar cell. This utilises the VHF_ PECVD approach, which provides a higher power density, essential for forming the microcrystalline phase. Efficiency enhancement can be achieved if there is a reflective layer between the two cells. This enables high current generation in a thin i layer top cell [64]. The emergence of the micromorph concept at the University of Neuchatel created widespread interest in the approach, with research groups and companies with a-Si:H experience (including Unisolar in the United States, Kaneka, Sharp, Mitsubishi, and Panasonic in Japan, and Bosch, T Solar, Schott Solar, Oerlikon, and Applied Materials in Europe) seeking to exploit their expertise. There were many others worldwide. The best stabilised micromorph-type cell had a stabilised efficiency of 12.7% ($1 \, cm^2$), while a triple-junction cell recorded an efficiency of 14.0% [57].

Despite this extensive research, the actual improvement in stabilised solar cell efficiencies between 1988 and 2016 was modest, and did not show the same year-on-year growth as similar technologies, with over 20% efficiency recorded for CdTe and CIGS, for example. Nevertheless, until recently, there was still significant manufacturing activity.

7.3.2.1 Amorphous Silicon Manufacturing

Although RCA laboratories demonstrated the first amorphous silicon solar cells, the lead in manufacturing came from Japan. In the late 1970s, demand for new consumer products such as digital watches and handheld calculators was booming. A range of crystalline silicon products had already been adopted as an alternative to batteries for use in these devices, and these made up a significant fraction of Solarex's production at the time [62]. Sanyo realised that the amorphous silicon solar cell would be cheaper to produce; even with only 3 or 4% efficiency, a few square centimetres would give adequate power. Amorphous silicon was preferred to crystalline as being easier to fabricate at the relevant size and giving a better output under fluorescent lighting [62]. Commercial production started in 1980, and by 1982 the first 1 MWp amorphous silicon manufacturing plant was operational [50]. A selection of Sanyo products is shown in Figure 3.13. Sharp Corporation also started amorphous silicon production at around this time, as did Kanegafuchi Chemical Industry Co. [65]. This contributed to amorphous silicon taking a major share of global production, as shown in Table 3.2. By 1999, Kaneka had the largest amorphous silicon plant in the world [66].

Amorphous silicon thus went from no production in 1980 to a global market share of 39% (13.9 Mwp) in 1988, almost exclusively in consumer products [50]. From this time on, however, while the consumer market grew only slowly, other market sectors were growing strongly, and while amorphous silicon production continued to grow, its share diminished.

Other companies were addressing the more commercial markets. Of particular note was Energy Conversion Devices (ECD) in the United States. As described in Chapter 3, ECD had been an early entrant into the amorphous silicon arena [53]. It was a small technology company based in Toledo, Ohio with expertise in amorphous materials. Its vision was to manufacture amorphous silicon through a high-volume process in a roll-to-roll machine using flexible substrates. Lacking its own funds for scale-up, ECD entered into a long series of collaborative ventures with major multinationals. The first was ARCO Solar in 1980, following a $25 million three-year investment [67]. This was not extended, and ECD subsequently supplied a machine to Sharp Corporation in Japan, although this does not seem to have been a commercial success and was wound up in 1987 [67]. ECD then formed a joint venture with Standard Oil of Ohio, establishing the Sovonics company in Cleveland to manufacture the ECD tandem cell. The activity was terminated after BP took full control of Standard Oil in 1988 and ECD requested additional funding to install a three-layer tandem manufacturing line. This was followed by a joint venture with Canon in Japan, leading to a 2 MWp p.a. plant, but this again ceased production [68]. In 2000, ECD formed a joint venture with the Belgian company Bekaert, setting up triple tandem-cell production in Auburn Hills, Michigan. This venture was short-lived, with Bekaert exiting in 2003 following significant losses in 2002 [69]. From then on, ECD ran its solar subsidiary United Solar Ovonic Corp (Uni-Solar) as a wholly owned operation, particularly addressing the commercial roofing market. This activity filed for bankruptcy in 2012 [70]. While ECD's concept had been good in roll-to-roll continuous manufacture, the use of the conducting stainless-steel substrate meant that the stainless steel had to be cut into strips, electrically isolated, and reconnected in order to make modules with useable voltages. This significantly increased costs as the monolithic connection used in glass superstrate modules could not be used. The company also had difficulty in increasing solar cell efficiency. While the best-stabilised module achieved 11.8% efficiency, typical production modules were only 8%, which was not competitive against either the First Solar thin-film modules at 12% or crystalline silicon at 16–20%.

While ECD and its venture partners were among the earliest to market, many other companies were also active in the arena. Sanyo continued to develop consumer projects, but only really produced demonstration products for larger-scale power plants

Figure 7.9 Sanyo's 'Solar Ark' with amorphous silicon panels (*Source:* Panasonic Corporation)

such as the 'Solar Ark' (Figure 7.9). Sharp continued its own production after terminating its relationship with ECD. In the United States, Chronar Corporation had its own proprietary technology, initially claiming to use disilane as the source gas rather than the monosilane used by all its competitors [71]. After establishing a 1 MWp manufacturing facility in New Jersey, Chronar branched out with a number of similar plants in the United Kingdom, France, the former Yugoslavia, Taiwan, and China [72]. Although the Chronar Corporation itself filed for bankruptcy in 1990, its technology continued through the Advanced Photovoltaic System company, which established a 10 MWp p.a manufacturing facility in Fairfield, California in 1992 [72]. ARCO Solar continued its research into amorphous silicon, and the company was sold to Siemens in 1990. Siemens also had its own amorphous silicon activity in Munich. The only other European manufacturer was Phototronics Solartechnik GmbH (PST), based in Putzbrunn near Munich. This started as a joint venture of Messerschmidtt-Bolkow-Blohm and Total Energie Development, using the 'Plasma-Box' technology developed at the Solems company in France [73]. Production at PST started in 1990 with a targeted 1 MWp p.a., and by 1998 was producing modules 1 × 0.6 m of typically 6.4% stabilised efficiency. The company passed through various ownerships before finally becoming part of the Schott Solar group, which closed in 2012.

While ARCO Solar, Siemens, and APS were active, Solarex launched a series of patent infringement cases against them in the United States and was successful in terminating their manufacturing activities [74].

The development by the original RCA group proceeded rather slowly. With the drying up of government funding in the early 1980s, RCA looked to sell off its technology,

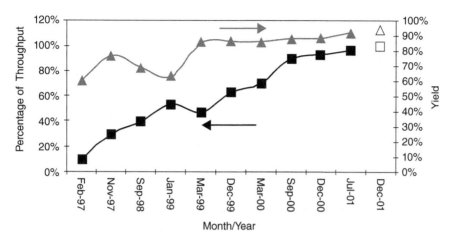

Figure 7.10 Throughput and yield of the Solarex a-Si module manufacturing plant [60] (*Source:* R.R. Arya and D.E. Carlson: Progress in Photovoltaics Science and Applications 10, 69-76. © 2002, John Wiley & Sons. Courtesy Wiley.)

and after some complex negotiations it was acquired by the American oil company AMOCO. AMOCO simultaneously acquired the established Solarex company and moved the amorphous silicon activity there, but at a separate facility in Newtown, Pennsylvania. Many key staff, including David Carlson, transferred there from RCA [75]. By 1993, the best modules were approaching 10% efficiency on 935 cm^2-aperture small-area modules [72]. Needing to expand production, Solarex became a joint venture between AMOCO and Enron in 1996 [75]. Large-scale manufacture started at a new facility in Toano, Virginia, making tandem modules of 0.77 m^2 with an efficiency around 8% and stabilising with a loss of up to 17% [60]. The production line was a one-off design by the Solarex team, and it took some time for a high yield and throughput to be established, as shown in Figure 7.10. It can be seen that it took nearly three years to go from start-up to 80% of rated output.

In 1998, BP merged with AMOCO, and in 1999 it bought out the Enron share to become the full owner of Solarex, which it ultimately combined with the existing BP Solar company. The product had a stabilised efficiency of around 6.5%, which was only half that of crystalline silicon. Therefore, the market for this technology was weak, and in November 2002 BP Solar took the decision to close the facility.

A number of other companies which started in amorphous silicon continued production by moving to the micromorph-type cell.

7.3.2.2 Manufacture of the Amorphous Silicon Microcrystalline Silicon Tandem Cell

A great stimulus to the manufacture of the 'micromorph' cell came from the fact that a number of established semiconductor equipment manufacturers were able to offer large-area inline PECVD deposition capability. This had developed from the need for precursor amorphous silicon layers for the emerging flat-plate display technology in the late 1990s [64]. The lead was given by the Oerlikon Group, which acquired the University of Neuchatel technology and adapted its KAI reactor systems to be able to process $1.4\,m^2$ micromorph cells. In 2008, Applied Materials also offered large-area capability, with the ability to coat a $5.7\,m^2$ area, usually comprising four panels of $1.1 \times 1.3\,m$. In addition, Anelva (Japan) introduced a Hot Wire CVD system capable of coating areas of $1.5 \times 0.85\,m$. These companies offered turnkey factory installations, and a number of manufacturing facilities were established worldwide. Applied Materials supplied around 12 customers, but it has now ceased offering production lines.

Oerlikon Solar was acquired by Tokyo Electric Solar (TEL Solar) in 2012 [76], having had a license for the technology since 2003. TEL Solar then reported the best performance to date for the micromorph module, with a stabilised 12.34% efficiency for an area of $1.43\,m^2$ [77]. However, TEL Solar terminated its manufacturing activity in 2014 [76].

Again, it was apparent that the relatively low efficiency of the product was unable to displace the established wafer crystalline product in the major markets.

7.3.3 Thin-Film Crystalline Silicon

Independently of the developments in amorphous silicon, it was recognised at an early stage that thin films of crystalline silicon on suitable substrates could produce solar cells with efficiencies as high as 27% in films as thin as $15\,\mu m$ [78]. This was supported by further studies which showed that good light trapping could increase the effective absorber width by a factor of 10, so that silicon films of $10-20\,\mu m$ could potentially have the same efficiency as a conventional $200\,\mu m$ wafer [79]. Realising this potential has proved a challenge from those early days up to the present time. Initially, experiments were directed toward depositing films on to graphite, ceramic or steel [80,81]. A 6.9%-efficient solar cell on steel with a $25\,\mu m$-thick film was reported in 1985 by the AstroPower company in collaboration with the University of Delaware and the target efficiency was reduced to a more realistic 20.5% [81]. Because of problems of contamination from the steel substrate, a ceramic barrier was introduced and increased in thickness until it eventually became free-standing and the steel was dispensed with [82]. Recrystallisation replaced liquid phase epitaxy as the method of fabrication of the silicon film. Early pilot production gave cells of $240\,cm^2$ area and 10.3% efficiency [83]. However, this was with a relatively thick silicon film of around $100\,\mu m$. Nevertheless, the capability to produce larger-area ceramic substrates increased, and by 1997, 15 cm-wide sheers were processed and cells of $675\,cm^2$ area gave 11.6% efficiency [84] – although the best laboratory cell was reported to give 16.6% efficiency on $1\,cm^2$ [85]. Development continued to scale up so that by 2002, a 15 MWp production line was established with 200 mm-wide ceramic substrates being processed and cells $208 \times 208\,cm^2$ being produced at a linear rate of 3 m/s, although efficiency remained relatively low at ~10% [86]. AstroPower also produced conventional silicon solar cells, but in 2004 it filed for chapter 11 bankruptcy and its assets were acquired by the General Electric Company (GE) [87]. GE quickly terminated silicon activities to focus on cadmium telluride development.

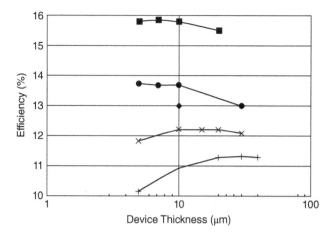

Figure 7.11 Efficiency of a crystalline silicon film with a minority carrier lifetime of 10 ns at different film thicknesses, showing the effects of different rear surface reflections [91] (*Source:* S.A. Edmiston: Progress in Photovoltaics Science and Applications 3, 333-350. © 1995, John Wiley & Sons. Courtesy Wiley.)

The difficulties of depositing thin-film silicon on various ceramic and steel substrates [88] and the advantages of glass as a substrate material have been highlighted. This was a particularly relevant topic in the early 1990s, as it was evident that the early promise of amorphous silicon for power applications was not being realised. The use of glass as a substrate introduces major constraints in the temperature that can be used and hence the method of deposition of the thin film. It is relatively easy to use PECVD at 700 °C on a graphite substrate to produce a 1 μm grain-size film with 5% efficiency [89], but in order to achieve a reasonable efficiency and cost, a grain size of 10 mm and a deposition temperature below 400 °C are required. Many techniques for depositing the film and creating large grains have been pursued [90]. However, an advantage of the thin-film silicon approach is than as the film thickness is low, the minority carrier diffusion length can also be low, so that high efficiency can be achieved with relatively poor material, as shown in Figure 7.11 [91]. It can be seen that nearly 16% efficiency may be achieved with a poor-quality thin silicon film with 80% rear surface Lambertian reflection. This assumes multiple stacks of p/n diodes. A target of 15% module efficiency was thought to be essential to overcome the limitations experienced with amorphous silicon modules.

While many research groups have been active, there has been only one serious attempt to develop a commercial thin-film silicon-on-glass product. This was a spin-out from the University of New South Wales, initially funded by the Australian utility Pacific Power. It led to the establishment of the Pacific Solar company in Botany Bay in 1995, with a pilot line started up in 1998 [92]. A novel device structure was used, as shown in Figure 7.12. The thin film was made by PECVD deposition on an amorphous silicon film stack of the active p+pn+ structure, with subsequent recrystallisation to

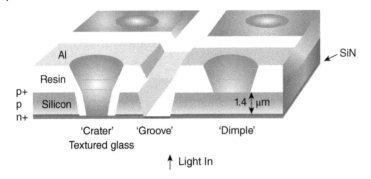

Figure 7.12 Interconnection scheme for crystalline silicon on glass [92] (*Source:* M.J. Keevers et al: EU PVSEC 2007, 22 th European Photovoltaic Solar Energy Conference and Exhibition, 1783- 1790. © 2007, WIP GmbH & Co Planungs-KG. Courtesy EUPVSEC.)

form crystalline silicon. The interconnection was achieved by scribing the active layer, coating it with an insulating resin, and creating a dimple and crater to allow simultaneous interconnection. There was sufficient lateral conductivity within the silicon that a TCO was not required. The active layer was only 2 µm thick. Initially, a module efficiency of just 2% was realised on a 980 cm^2 module, but this rose to just over 7% by 2001. In 2004, the activity in Australia was terminated and re-established as Crystalline Silicon on Glass Solar (CSG) in Thalheim, Germany, part-funded by Q Cells [93]. A 9% small-area module efficiency was reported in 2004 [94], with 10% efficiency in a submodule of 94 cm^2 [95]. By 2009, low-volume production had begun and more than 6 MWp had been deployed, with production running at an annual rate of 13 Wp. However, efficiency was only 6.6% (active area) for a 1.38 m^2 module [96] – much lower than for other thin-film modules. CSG ceased production in 2012 [97].

7.3.4 Copper Indium Gallium Diselenide (CIGS)

The original form of this semiconductor was copper indium diselenide (CIS), which was reported in 1981 as one of the four emerging thin-film technologies giving a laboratory efficiency of over 10% [41]. The impetus for CIS was the groundbreaking development in 1975 of a method for growing epitaxial cadmium sulphide on a single crystal of CIS and producing solar cells with an AM1 efficiency of over 12% (0.07 cm^2) [98]. This broke the 10% efficiency barrier, which no other material than silicon had achieved at that time, and provided a spur for further development, which has continued to the present day. Attention soon turned to the deposition of thin-film multicrystalline CIS, and a 6.6%-efficiency cell (1.2 cm^2) was reported by SERI (NREL) in 1976 [99]. Early theoretical analysis showed that the CIS technology could achieve an efficiency of 26% [100]. Improvement in solar cell efficiency came rapidly, and Boeing achieved a 9.4% (1 cm^2) cell by 1981 [101] and broke the 10% barrier in 1982 [41].

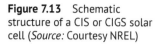

Figure 7.13 Schematic structure of a CIS or CIGS solar cell (*Source:* Courtesy NREL)

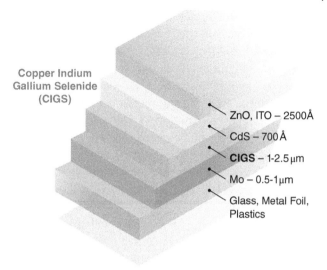

Copper Indium
Gallium Selenide
(CIGS)

ZnO, ITO – 2500Å

CdS – 700Å

CIGS – 1-2.5 µm

Mo – 0.5-1µm

Glass, Metal Foil,
Plastics

The structure of a typical cell is shown in Figure 7.13. The film is deposited on to a substrate, which is usually nonconducting and in the early days was either glass or alumina [101]. A metal film is then deposited as the back contact; this quickly became standardised as molybdenum. The active layers are deposited on top, a CdS layer is added to form the n type region, and finally a TCO is added to form the top contact. The best early results came from depositing the CIS by co-evaporation of the elemental sources. Control of the stoichiometry of the CIS film was critical to achieving a high efficiency [101]. Careful geometric arrangement of the sources was required to get good compositional control, and an overpressure of selenium – the most volatile species – was essential. The CIS film is always p type, presumably due to copper vacancies. Intentional p type doping by selected impurities is not achievable for CIS [102]. The CdS was deposited by evaporation in the early stages of development, but chemical bath deposition gives the best solar cell results [102].

Other techniques can also be used, such as sputtering [103] and electrodeposition [104], although the results are inferior to co-evaporation. One other good technique, the two-stage process, involves sputtering the metallic elements either as single elements or as selenides and converting them to CIS by heat treatment in a hydrogen selenide atmosphere. Efficiency results were comparable to the co-evaporated route, with 6.1% efficiency on a 4 cm² cell area [105].

A further advantage of CIS as a solar cell material is that the devices produced are intrinsically stable and do not suffer the inherent light-induced degradation seen with copper sulphide and amorphous silicon cells [101]. CIS is also radiation-hard, indicating its potential as a lightweight low-cost material for use in space [106].

The main thrust of research through the 1980s was toward increasing the efficiency of the CIS solar cell. It was quickly recognised that the bandgap in CIS at 1.1 eV, like that in silicon, was not optimal. There had been an early attempt to make a solar cell in copper gallium diselenide (CGD) with a bandgap of 1.68 eV [107]. Copper Schottky diodes were made on single crystals of CGD and gave a solar cell efficiency around 4%, as measured in direct sunlight. A later result gave a 3.7% efficiency on a small-area cell [108]. A more detailed study aimed at broadening the bandgap was carried out using both gallium to substitute for some of the indium and tellurium to substitute for some of the selenium, using sputter deposition [109]. The highest Voc was given by a $CuGa_2Se_{0.9}Te_{1.1}$ cell (bandgap ~1.4 eV) of 500 mV, as compared to the CIS reference of 310 mV; theoretical studies using the data generated showed that a bandgap of 1.45 eV gave the highest potential efficiency, approaching 26%. A CGD cell achieved 5% efficiency and it was shown that the addition of gallium to CIS broadened the bandgap systematically [110]. The first successful CIGS cell was reported in 1985 by Boeing, with a 7.2% efficiency [111].

CIGS rapidly became the material of choice for future research, with NREL leading the way and achieving a small-area efficiency of 13.3% by 1993 [112]. NREL was particularly successful in developing the three-stage process for deposition of the CIGS layer by evaporation. With a background constant selenium evaporation, indium and gallium in the appropriate proportion were deposited at a lower temperature of 300–400 °C. The substrate temperature was then raised to 500–600 °C and copper was deposited, followed by further indium and gallium. The result was a segregation of gallium toward the front and rear surfaces, giving a graded bandgap with minimal recombination at both surfaces of the absorber layer and increasing efficiency. This led to a solar cell efficiency of 15.9% in 1994 [113]. New impetus was given to the technology by European research institutes, notably in Germany and Sweden, with 16.9% active-area small-area cells being reported in 1993 for co-evaporated CIGS [114]. By 2000, efficiencies of up to 19% were reported for small-area CIGS cells [115] and the factors enhancing efficiency in CIGS were becoming better understood. In particular, the incorporation of sodium into the film had beneficial effects in increasing p type conductivity, promoting large grain growth, and stabilising the formation of the CIGS phase. The use of chemical bath deposition (CBD) for the deposition of CdS also brought benefits compared to the evaporated CdS in terms of better coverage of the CIGS, stabilisation of the crystallographic surface phases, and minimisation of recombination at the heterojunction. With further optimisation of these parameters, a 19.9%-efficient solar cell (area 0.419 cm^2) was reported by NREL in 2008 [116]. The 20% efficiency level for CIGS was achieved at ZSW in 2010 for a 0.5 cm^2 solar cell [117]. The world record for a CIGS cell was also extended by ZSW, which reported a 22.6% efficiency (0.5 cm^2) in 2016 with improved alkaline post-treatment contributing to the improvement. The use of potassium instead of sodium for the alkali treatment was beneficial [118].

As well as improving world record efficiencies, research has been directed toward other areas of concern in the CIGS technology. There are many environmental issues regarding the use of cadmium in solar cells, and much work has been done to find alternative cadmium-free buffer layers with compounds such as $Zn(O,S)$ and In_xS_y.

Integrating these approaches has led to the current world record solar cell of 23.35% efficiency (1 cm^2) created by Solar Frontier using the potassium fluoride post-deposition treatment and a wide-bandgap zinc magnesium oxide window layer [119]. This cell is also cadmium-free.

Concerns have been raised about the availability of indium if production levels reach the tens of gigawatts level of CIGS [120]. This has stimulated research into very-thin CIGS absorber layers. The record solar cells just mentioned typically had 2.5 µm-thick absorbers. With optimisation of the gallium content, it is possible to make cells with submicron absorber layers that can achieve 90% of the efficiency of a standard-thickness cell [121].

In addition to glass substrates, CIGS can be deposited on polyimide foils to form flexible modules. Using this approach, 20% efficiency has been achieved on small areas [122].

7.3.4.1 CIGS Manufacturing

While, as just described, there is a long history of the achievement of high efficiencies in CIGS, manufacturing has taken off much more slowly than for cadmium telluride modules. Figure 7.14 compares annual shipments of CIGS with those of other thin-film technologies. It can be seen that CIGS had negligible production before 2008, with output lagging behind cadmium telluride and keeping pace with amorphous silicon until the latter declined in 2014. This may in part be due to the difficulties in the scale-up of the CIGS technology. To quote the distinguished CIGS scientist H.W. Schock, 'manufacturability has always been an issue' [115]. Various attempts have occurred. The first was by ARCO Solar, which in 1985 demonstrated a four-terminal a-Si/CIS

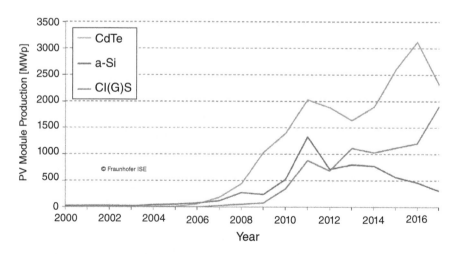

Figure 7.14 Annual shipments of thin-film photovoltaic modules by technology (*Source:* PHOTOVOLTAICS REPORT, Prepared by Fraunhofer Institute for Solar Energy Systems, ISE with support of PSE GmbH, Freiburg, 18 May 2019. © 2019, Fraunhofer ISE.)

mechanically stacked module [123] that achieved 13% efficiency in a small area (4 cm^2), with potential for over 20%, but had just 7% efficiency in larger-area devices [124]. This approach was dropped when ARCO Solar lost its rights to the a-Si technology, as described earlier. Ownership of the company then passed to Siemens, which continued with CIS alone.

In 1993, the 10% barrier for a large-area module was broken when NREL confirmed 10.5% efficiency for an area of 3890 cm^2 [125] with good stability in outdoor testing. Nevertheless commercial activity remained very much at the pilot- and demonstration-project level [126]. The situation had not changed very much by 2000, when only Siemens had a commercial-size module (1×4 foot), exhibiting 12.1% best efficiency [115]. The company then went through a number of corporate changes. Focus on CIGS research moved from the United States to Germany, where Siemens developed a rapid thermal process for forming the absorber layer after vacuum deposition of the elemental stacked layers. The Siemens Solar activity was sold to the Shell Solar in 2002. Then, Shell and the St Gobain group formed a joint venture, Avancis, to commercialise the technology. In 2009, Avancis became wholly owned by St Gobain, and a joint venture with Hyundai led to the establishment of a production facility in South Korea in 2012. In 2010, Avancis was producing modules in its PowerMax range up to 130 Wp (~1 m^2). Efficiency improvement came from the RTP process and the use of St Gobain Securit Albarino ARC glass. Sulphur was added to increase the bandgap to 1.53 eV. Smaller-area (668 cm^2) modules then demonstrated a 15.1% efficiency [127]. In 2014, Avancis was acquired by the Chinese group CNBM, and in 2017 the first modules were produced at the Bengbu (China) plant, which is now scaling up to a 1.5 GWp annual production capacity [128]. The highest-power product currently being produced is the POWERMAX module at 150 Wp (14.2% efficiency).

The work at the leading ZSW group went into pilot production with Wuerth Solar in 2002. ZSW did material development on small-area devices, while Wuerth scaled up the technology on 0.7 m^2 modules, with the champion module giving 74 Wp and 12.5% aperture efficiency [129]. In 2010, the Manz group took over the activity from Wuerth, still with ZSW collaboration. In 2015, Manz reported a record 16%-efficiency module, although its commercial activity is centred on the sale of CIGS production lines [130].

The Nanosolar company raised substantial venture capital funding (over $500 million) in the United States to set up large-scale manufacturing of printable CIGS using specially formulated inks with nanocrystalline particles. It targeted a 430 MWp annual production, with plants in the United States and Germany. While small-area cells gave a 17.1% efficiency, module efficiencies were below 10% in 2011. The company closed in 2013 with less than 50 MWp shipped [131].

The CIGS development at Uppsala University was initially commercialised by Solibro in Germany, which started production with a 100 MWp p.a. line at Thalheim. Solibro was taken over by Hanergy in 2012, and Solibro GmbH and its associate company Solibro HiTech GmbH began insolvency proceedings at the end of 2019 [132]. A 21% small-area cell was reported in 2017, with a 15.6% (1 m^2) module [132]. Hanergy has continued its activities through the MiaSolé subsidiary in China, producing flexible CIGS modules, the best of which measures at 18.6% efficiency (1.08 m^2) [133].

The other major CIGS manufacturer is Solar Frontier. The original development was carried out over many years by the Showa Shell company in Japan. It reported a 13.5% small-area cell made by the selenidisation process in 1994 [134]. By 2002, it was achieving 12.2% efficiency on a 900 cm^2 module [135]. Solar Frontier now has three manufacturing plants with a total capacity of 4 GWp p.a. and produces a standard module of 185Wp rated power with 15% efficiency (1.22 m^2).

CIGS manufacturing is at an early stage, lagging behind cadmium telluride. It remains to be seen if this technology can capture a significant market share in the long term.

7.3.5 Cadmium Telluride

As shown in Figure 7.14, cadmium telluride (CdTe) has become the leading production technology for thin-film modules. It has a number of advantages over CIGS in that the stoichiometry is much easier to control as it is the only compound that cadmium and tellurium form; it is both congruently melting and evaporating, which makes the deposition of the thin film simple and allows a range of manufacturing techniques to be employed; and it has a near ideal bandgap for a single-junction solar cell at 1.4 eV. It was one of the four thin-film technologies identified in 1982 as breaking the 10% cell efficiency barrier [41].

One of the earliest papers in 1963 reported a 6%-efficient thin-film solar cell heterojunction of p type copper telluride on n type CdTe [136]. Notable progress was made in 1965, when in its search for alternatives to silicon and copper sulphide for use in space cells, the Air Force Aero Propulsion Laboratory (United States) fabricated an n type CdTe cell. This was done by depositing n type CdTe on to a CdS-coated molybdenum substrate and then forming a heterojunction with a p type copper telluride formed via ion exchange by dipping the CdTe into a copper solution. A gold grid was evaporated to form the front contact. An efficiency of 5% was recorded in natural sunlight at 85 mW/cm^2 on a relatively large area of 56 cm^2 [137]. The robustness of CdTe to different deposition methods and the ability to form a range of junctions – either homojunctions, heterojunctions, or MIS devices – led to a wide range of approaches [138]. The ultimately successful pCdTe/nCdS heterojunction was first demonstrated in 1972 [139]. CdTe was deposited at around 650 °C and then CdS at around 200 °C to form an abrupt junction. A solar cell efficiency of around 5% was achieved. By 1982, a number of approaches had reached some degree of maturity. The 10% barrier was broken using the CdTe/CdS structure [140]. The configuration of the cell has become the standard for subsequent development. The substrate was an indium tin oxide-coated soda lime glass. The CdS and CdTe were deposited at high temperature by closed space sublimation (CSS) in a partial vacuum without a transporting gas. The CdS was deposited first on to the n type ITO at 550 °C and then the CdTe at 600 °C. The rear contact was evaporated gold. A 10.5% efficiency was measured for a $0.1\,cm^2$ solar cell with approximate AM2 spectrum and $75\,mW/cm^2$. At the same conference, a CdTe/CdS cell was made by electrodeposition of the semiconductor layers, and small-area cells of $4.2\,cm^2$ area achieved 7% efficiency [141]. Screen printing technology also showed promise, as an all screen-printed CdTe/CdS cell on a borosilicate glass substrate gave an AM1.5 efficiency of 9.0% ($50\,cm^2$) [142]. A year previously, spray pyrolysis had been used successfully to deposit thin-film CdS and CdTe, forming a 4%-efficient solar cell ($0.024\,cm^2$ AM1.0) [143]. These four approaches formed the basis for the ensuing device research over the next two decades.

Although a number of research approaches were followed, it became possible to describe a generic technology by 1995 [138]. A glass superstrate with either an ITO or tin oxide TCO is used. A CdS film is deposited by a range of techniques and usually heat-treated with $CdCl_2$ to increase grain size and decrease defect density. The CdTe absorber is deposited by a range of techniques in thicknesses varying from 1.5 to $15.0\,\mu m$, the latter being used for screen-printed films. A further $CdCl_2$ treatment is then applied. Some interdiffusion of tellurium and sulphur usually occurs at this stage. The rear contact is a challenge. It is typically a two-layer structure in which either chemical etching is used to produce a tellurium-rich p+ surface or a p+ semiconductor such as ZnTe, HgTe, or PbTe is applied. A number of metals or carbon may be used as the final metallisation. Back in 1982, Tyan noted that critical areas for improvement were the limitations of the CdS window layer, the quality of the CdTe absorber, and the achievement of a low-resistance contact to p type CdTe. These topics have remained the focus of much subsequent development.

By 1993, the best cell efficiency had increased to 15.8% ($1.05\,cm^2$, verified by NREL), using a CBD CdS and CSS CdTe [144]. CdTe modules were also demonstrating good stability, showing less than 1% relative efficiency degradation after $16\,000$ hours on external test [145]. By 2000, a best laboratory cell of 16% had been reported at a number of laboratories [146] and good progress was being made with commercial-size modules. Progress in high efficiency was relatively slow in the period from 2001 to 2011, when the best reported cell efficiency was 16.7%; only in 2012 did a significant increase occur, with GE reporting an 18.3% ($1\,cm^2$)-efficient solar cell [147]. Best laboratory cell efficiencies then continued to rise up to the present record of 22.10% ($1.06\,cm^2$), reported by First Solar in 2016 [148]. This was achieved through the use of improved CdS window layers, a graded bandgap in the absorber with a narrower bandgap CdSeTe layer near the buffer layer, and a ZnTe/metal rear contact.

7.3.5.1 Cadmium Telluride Manufacturing

Figure 7.14 shows the rise of CdTe commercial module production. This is almost entirely due to the work of one company, First Solar. Through the 1980s, as already described, a number of companies were active in pilot-scale production, but one by one they fell aside by the start of the 2000s. In 1995, five major operations were identified [138]. One of the earliest commercial companies was Photon Energy, which was taken over by the Coors Company in 1991 to become Golden Photon. By 1995, it had achieved 9.1% efficiency on a $0.34\,m^2$ module using the spray pyrolysis technology [146]. However, the operation was terminated in 1997. The screen-printing technology pioneered by Panasonic (Matsushita) was then dropped in favour of CSS, but Panasonic did not commercialise this technology. In 2008, Panasonic acquired the Sanyo Electric company, in part so that it could exploit Sanyo's leading silicon HIT technology [149], which became its lead solar technology. BP Solar acquired the electrodeposition technology from Monosolar in 1984 and proceeded to develop it for large-area modules [145]. It established a production facility on the former APS factory in Fairfield, California, where it was able to manufacture large-area modules of 10% efficiency, with a record 11.0% efficiency for a $1.2\,m^2$ module in 2002 [150]. However, the company took the view that the rapid growth of the photovoltaics market anticipated at the time would be best addressed by expansion of its silicon solar cell technology, so it terminated its CdTe activity. At around the same time, in Germany, following on from the development of CdTe at the Battelle Institute in Frankfurt, the ANTEC company was set up to manufacture CdTe modules with an anticipated throughput of $100\,000\,m^2$ p.a. based on CSS technology. The line was operational by the end of 2001, producing 60 x120 cm modules of 7% efficiency [151]. However, the product was not successful in the highly competitive German market and the activity became insolvent in 2002 [152].

The most successful company to date is First Solar. It was originally set up as Solar Cells Inc. in 1990 with a mission to find a successful thin-film solar technology. After initial interest in a-Si, the company focussed on vapour transport for CdTe manufacture. In this approach, a heated vapour stream transports metallic cadmium and tellurium from an evaporation source to a moving substrate, where CdTe is deposited at temperatures of 450–600 °C [153]. In 1993, SCI was able to produce a 7.4% aperture efficiency on a 60×120 cm module in a 100 kWp pilot facility. The advantages of the vapour transport method – which is a variant on the CSS technique in which the source-to-substrate distance is increased – include less critical operating parameters and faster deposition [154]. Following a new round of investment in 1999, SCI became First Solar and continued expansion of its activities. In 2002, a facility producing 3000 modules per month (~2 MWp p.a.) was established in Perrysburg, Ohio and achieved 8.2% efficiency ($0.72\,m^2$), with a champion module of 10.1% efficiency [155]. By 2005, this had expanded to 25 MWp and module efficiency was steadily rising. The timing was good as there was a very strong market in Europe, where demand was outstripping supply and there was a shortage of silicon modules, which made installers open to using other technologies that they might not otherwise have considered. In 2007, First Solar started production in Malaysia, and by 2009 it had achieved 1 GWp capacity, making it the second largest photovoltaic manufacturer in the world at the time. By 2013, module costs had fallen to $0.68/ Wp [156], and by 2016 module efficiency had risen to 16%. In 2010, the company made a strategic shift from being a module supplier to installing large grid-connected

Figure 7.15 Topaz Solar Farm, Sacramento, CA with 550 MWp of First Solar modules (*Source:* Solar Panels at Topaz Sola, Pacific South West Utility, 2012. *Source:* Sarah Swenty/USFWS, https://commons.wikimedia.org/wiki/File:Solar_Panels_at_Topaz_Solar_5_(8159036498).jpg. Licensed under CC BY 2.0)

systems, particularly in the United States (see Figure 7.15). This allowed the company to ensure that the locations of its products were clearly identified and end-of-life disposal and recycling could be controlled as the means to overcoming the known environmental hazards of cadmium-containing products [157]. First Solar has over 3 GWp of manufacturing capacity, and in April 2018 it announced a 1.2 GWp expansion of its series 6 manufacturing in the United States [158]. The series 6 module is very large-area (2.4 m^2), making it suitable for the large solar farm market. Currently, the module efficiency is 18% [159].

While First Solar has made significant progress and holds a well-established market position, its global market share continues to decline, as shown in Figure 4.1.

7.3.6 Dye-Sensitised Solar Cells

The interaction of light with an electrolyte to create a voltage between electrodes has a long history, dating back to Edmond Becquerel in 1839 [160]. However, it was not until the 1960s that the potential for photoelectrochemical cells became apparent. Photoelectrochemical cells typically consist of a semiconductor photoelectrode, a redox electrolyte, and a counterelectrode. Oxide semiconductors are generally stable in contact with electrolytes but have too wide a bandgap to capture the solar spectrum, so a dye is used to inject electrons into them [161]. Initial experiments were made with zinc oxide sensitised with organic dyes [162]. These devices were of low efficiency as

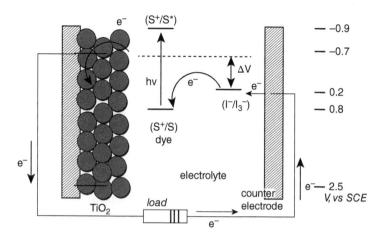

Figure 7.16 Mechanism for a DSSC [164] (*Source:* M. Gratzel: Progress in Photovoltaics Science and Applications 8, 171-185. © 2000, John Wiley & Sons. Courtesy Wiley)

the low surface area of the semiconductor gave inadequate light collection. The major advance came in 1991 [163], when O'Regan and Grätzel reported a 7%-efficient solar cell using a ruthenium biphenyl dye absorbed on to 10 μm of nanostructured titanium oxide with an iodine redox electrolyte. This was seen as having high potential for low cost, as TiO_2 was a cheap and relatively abundant source material which did not require the complex processing of the other potential semiconductor materials conventionally used in photovoltaic cells.

The operation of a dye-sensitised solar cell (DSSC) is shown in Figure 7.16. A photon is absorbed in the dye and an electron is immediately injected into the TiO_2. The TiO_2 is deposited on a TCO, which collects the electron and delivers it to the external load. The electron then returns to the counterelectrode and, through the redox reaction in the electrolyte, reverts the now positively charged cell to its neutral state.

The initial announcement stimulated a worldwide research effort into DSSCs. Improvements were made in the structuring of the TiO_2 and the composition of the dye. However, progress was relatively slow. A 10%-efficiency cell was not independently confirmed until 2000 [164]. A major issue was sealing the liquid electrolyte so that DSSCs could be used in the wide range of climatic conditions experienced by photovoltaic cells. In addition, the stability of the dyes under long-term exposure was questionable [161]. Efficiencies have remained low compared to the competing silicon and thin-film technologies, ranging from 18 to 20%. The best independently verified DSSC efficiency is 11.9% for a 1 cm^2 laboratory cell, with 8.8% for a 398.8 cm^2 submodule [148]. No significant commercialisation has occurred.

It is probable that the significance of the DSSC is that it was the precursor to the emerging perovskite technology described in 7.3.8.

7.3.7 Polymer (Organic) Solar Cells

The heavy investment in research in DSSCs indicates the importance that was given to finding alternatives to the conventional semiconductors. While a photovoltaic effect had been observed in organic materials in the 1950s, it did not become a topic of research until the late 1990s [165]. In the display field, organic light-emitting diodes (OLEDs) achieved success based in part on research work at Cambridge into new light-emitting organic materials [166]. It seemed a logical step to look at organic semiconductors for solar cells, and from 2000 on many research groups were active worldwide. OPVs have different properties to inorganic ones, and this creates many challenges. The photo-excited electron remains bound to the hole as an exciton and only separates at an energetically favourable interface. For this separation to occur, two types of polymer are brought into contact, one being an electron donor and the other an electron acceptor. The diffusion length of excitons is only a few tens of nanometres [165], but as the OPV is highly absorbent this is all that is required to harvest the incoming light. This has given rise to two structures. The first contains very thin-film small-molecule polymers, which are typically deposited by vacuum deposition [167]. The second contains large-molecule polymers, which are co-deposited so that an interpenetrating structure of the two conduction types is achieved, as shown in Figure 7.17. This is called the bulk heterojunction device and is the main device on which research is performed.

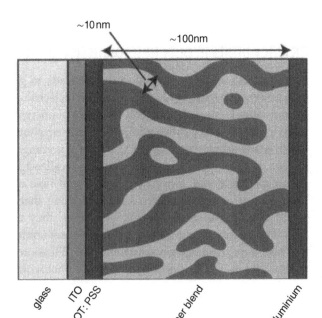

Figure 7.17 Schematic of a bulk heterojunction OPV, showing the interpenetrating polymers connecting to the appropriate electrodes [168] (*Source:* Greenham, N. C. Polymer solar cells. Philosophical Transactions of the Royal Society A: Mathematical, Physical and Engineering Sciences, 371(1996), 20110414–20110414. Courtesy The Royal Society.)

The attraction of OPVs is that they represent a potentially low-cost technique, as very little absorber material is required and the films can be deposited from solution by a range of methods, including screen printing, spraying, inkjet printing, and doctor blading [165]. These methods are inherently scaleable and easily patterned. The process is usually carried out at room temperature, keeping production costs potentially very low. There are numerous reviews of OPV technology (e.g. [165,168]), so only a short summary will be given here.

Figure 7.17 illustrates a typical cell structure. The absorber layer is composed of the interpenetrating electron donors and acceptors. These are nanostructured so that there is a large area between them, allowing exciton separation and providing a short distance for the exciton to travel before charge separation. The deposition of the polymer blend has to ensure that the majority of each carrier type contacts the appropriate electrode. The electron donor most frequently used in early developments was P3HT (regio-regular poly(3-hexylthiophene)), while the electron acceptor most commonly chosen was soluble fullerene, $PC_{60}BM$ (6,6-phenyl-C-61-butyric acid methyl ester). The conducting transparent polymer PEDOT:PSS (Poly(3,4-ethylenedioxythiophene) polystyrene sulphonate) was used to facilitate contact to the inorganic TCO contact, as this had a work function intermediate between the TCO and the polymer. A variety of materials have been investigated for use as the hole acceptor.

A 7%-efficient OPV cell was reported in 2010 [169], while the most recent efficiency recorded in 2020 was 13.45% ($1 \, cm^2$) [170]. No significant manufacturing has occurred for OPV devices. A major issue to be overcome is the long-term stability, as the OPV cell is particularly sensitive to water vapour. From 2015 on, the focus shifted from copolymer/fullerene materials to metalorganic perovskite ones.

7.3.8 Perovskite Solar Cells

This is the most recent solar cell technology to have demonstrated the potential for a low cost and high efficiency comparable to those of crystalline silicon. The perovskite solar cell (PVK) technology has shown very rapid progress, going from a reported 10% efficiency in 2012 to 21.6% ($1 \, cm^2$) in 2020 [170]. The remarkable rise of best solar cell efficiencies for PVKs is shown in Figure 7.18. This demonstrates how in the space of a few years, the best PVK efficiencies have caught up with the established CdTe and CIGS ones and are approaching those for silicon and III–Vs.

The PVK approach arose from the need to find a solid-state electrolyte for DSSCs, as already described. There were some early experiments with PVK and liquid electrolytes, although these had poor stability [171]. The breakthrough came when the PVK was used alone. A solar cell efficiency approaching 10% was found when a mixed halide PVK replaced the electrolyte in a conventional DSSC [172], but the transforming result was the observation at the same time that a 10.9%-efficient solar cell could be made with insulating alumina replacing the TiO_2 in the DSSC, indicating that the PVK alone could be the active solar cell material [173]. The metal organic compound was

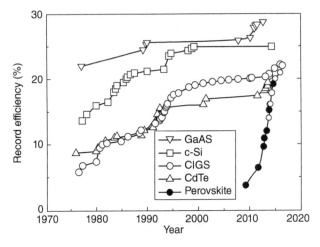

Figure 7.18 Best solar cell efficiencies with time for different thin-film technologies [171] (*Source:* S.D. Stranks and H.J. Snaith : "PV Solar Energy-from fundamentals to applications" ed A. Reinders, P. Verlinden, W. van Sark, A.Freundlich, Pub J Wiley, 275-291. © 2017, John Wiley & Sons. Courtesy Wiley)

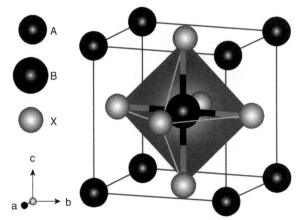

Figure 7.19 Perovskite crystallographic structure, where A is a short-chain organic cation, B is a metal cation, and X is a halide [171] (*Source:* S.D. Stranks and H.J. Snaith : "PV Solar Energy-from fundamentals to applications" ed A. Reinders, P. Verlinden, W. van Sark, A.Freundlich, Pub J Wiley, 275-291. © 2017, John Wiley & Sons.)

obtained by crystallising from solution the organic halide with the metal halide to form the crystallographic perovskite structure ABX_3, where A is a short-chain organic cation such as methylammonium, B is a metal cation (usually lead), and X is the halide (normally iodine). The perovskite structure is shown in Figure 7.19.

Most research to date has centred around the methylammonium lead iodide complex. All of the components can be replaced by other elements. The methylammonium can be replaced by formamidinium (FA), a larger cation, which may give better thermal stability [171]. The absorption edge can be modified using chlorine and bromine instead of iodine. Other elements can replace lead, although tin-based PVKs have a much greater tendency to oxidation [171].

A typical PVK consists of a fluorine-doped tin oxide-coated glass with a dense TiO_2 blocking layer, a very thin mesoporous TiO_x coated with the intrinsic PVK, a hole transport layer (normally Spiro-OMETAD), and finally a metal contact.

Figure 7.20 External quantum efficiency (EQE) of PVK and silicon solar cells [174] (Courtesy Wiley)

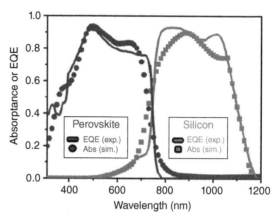

The highly ordered perovskite structure gives the compound its good electronic properties. PVKs have the highest ratio of Voc to the bandgap of any of the common solar materials, namely 1.1 V for MAPbI$_3$ with a bandgap of 1.55 eV. The binding energy for the exciton is only a few meV so that free carrier transport can occur immediately after photon absorption. The electron and hole diffusion lengths are then long compared to the film thickness, giving good current generation [171].

The PVK technology is relatively new but the progress is achieving high efficiency has been remarkable. The current record efficiency for a small-area PVK cell is 25.2% (0.094 cm^2) [170]. At this early stage of development, the best efficiency falls strongly with area, and the best submodule to date has an efficiency of 16.1% (802 cm^2) [170]. Many issues are yet to be resolved, including in long-term stability and the acceptability of lead as a significant component, as well as the technology for large-area deposition and production costs. Rather than being introduced as a totally disruptive technology, its future may be as an add-on to conventional silicon technology. As a polymer, there are no lattice mismatch issues in applying a PVK to silicon, and the potential for efficiencies up to 30% using this tandem has been identified [174].

The good complementarity of silicon and PVK is shown in Figure 7.20, with an early confirmation result of 23.6% efficiency for a tandem cell in 2017 [175]. The potential for PVK tandem cells on silicon with more recent results is discussed more fully in Chapter 8.

7.4 Concentrator Technologies

In the early 1980s, three approaches to lowering the cost of photovoltaic solar systems were identified in the United States. The first was to lower the cost of silicon manufacture through the LSSA programme, as described in Chapter 2. The second was to develop new thin-film solar cells, as described earlier in this chapter. The third was to apply optical concentrations of sunlight on to small-area solar cells using lenses or mirrors. The advantages of the use of concentrators are clear. Instead of presenting a large collecting surface of solar cells to incoming sunlight, it is more cost-effective to use low-cost lenses or mirrors to direct the light on to small-area cells, thereby replacing large areas of expensive semiconductor with plastic or metal. In addition, the sun will necessarily be tracked, giving a more uniform energy production throughout the day as compared to a fixed flat-plate array. These arguments could be easily justified in the 1980s, when photovoltaic system prices were $8/Wp and the cell cost was $5/Wp. If a 100× concentrator could be used then the cell cost would fall to $0.05/Wp, and if the additional optics and tracker were to cost $1/Wp then a system price of $4.05/Wp would be achievable and a significant advance would be made in lowering the cost and enhancing the acceptability of photovoltaics.

The development of concentrator systems has a long history, as reviewed by Swanson [176], with proof-of-concept studies going back to the 1960s. Through the 1980s, a number of demonstration projects were started [177], generally using silicon solar cells and point-focus Fresnel lens systems. Efficiencies were low at around 10%. No commercialisation beyond the demonstration phase occurred, and in some cases there were major system failures [178]. The reasons for the low take-up before 2005 have been reviewed by Bruton [178], who found the chief issue was the nature of the market up to that date: most installations were rooftop and small commercial systems of less than 5 kWp, while concentrators did not become cost-effective until systems greater than 50 kWp were required. Although companies like Entech were able to demonstrate good performance [179], outperforming flat-plate systems, there was not sufficient demand to allow them to achieve a scale of manufacture that made concentrator systems cost-effective.

A new surge of activity occurred in the mid-2000s as total demand grew to exceed supply and large grid-connected plants at the 10 MWp level became common in Europe, especially Germany. A new generation of triple-junction III–V cells became available from suppliers such as Spectrolab and Emcore and a number of start-up companies emerged. Before 2005, the leading concentrator company had been Amonix. Amonix was founded in 1989 to exploit the concentrator concept promoted by EPRI for a high-efficiency rear-contact silicon cell with a point-focus Fresnel lens. By 1996, it had produced three 20 kWp demonstration systems at 250× with 26%-efficient silicon solar cells [180]. It continued to deploy systems, particularly at the Arizona Public Service test site in Phoenix. However, by 2004, it was shown that a tracked flat silicon array gave a 20% higher kWhr/kWp yield than the Amonix system [181]. At this time, Amonix switched to 500× concentrator systems using the triple-junction III–V cells. It carried out further development and had installed 70 MWp worldwide by 2014, when manufacturing ceased. Another leading company was Concentrix, which was founded in 2005 to exploit the concentrator and solar cell technology developed at the Fraunhofer Institute for Solar Energy [182]. An example of the Concentrix system is shown in Figure 7.21. Using a triple-junction Ge/GaInAs/GaInP solar cell, 500× point-focus systems with 27% efficiency were developed. Various projects followed, and a number of 10 MWp-size systems were constructed. The company was sold to Soitec in 2010. Despite achieving a world record in 2016 for what is still the highest-efficiency

Figure 7.21 Flatcon 500× receiver at FhG.ISE [182] (*Source:* H.Lerchenmuller et al: 4th International conference on solar concentration for the generation of electricity or hydrogen (2007) 225–228. Courtesy FhG.ISE)

concentrator cell to date, with 46% efficiency at 508× using a four-layer wafer-bonded tandem cell in 2016 [148], Soitec terminated the business [183]. This left no global producer of concentrating systems.

The problem with concentration was that the economic model which had made it attractive in the 1980s was no longer valid. Along with the growth in the market, silicon module prices had fallen dramatically to less than $0.6/Wp and system costs were around $1/Wp. This meant that the savings in cell costs through concentration were not able to pay for the additional complications of module assembly using many small cells, lenses, and trackers. An additional factor was that concentrators could only collect direct global insolation – typically 85% of the total insolation. Optical losses in the lens were at best 20%, leaving the actual insolation at the cell less than 70% of that in the flat-plate case. This meant concentrator systems were only suitable for use in extremely high-DNI areas, which are generally remote. All these factors added up to make concentrators with III–V cells uneconomic despite their very high efficiency – hence the current negligible global market.

7.5 Summary

This chapter has described a range of technologies which have the potential to displace the conventional route to the manufacture of silicon modules by polysilicon production, crystallisation, wafering, solar cell fabrication, cell interconnection, and module encapsulation. Over the past four decades, many alternative technologies have been tested. Those reviewed in this chapter represent only a small cross-section of the total, as the focus has been on those which achieved some degree of commercialisation. This is surprising, as in many instances the activities were well financed over long periods of time and carried out by highly competent research organisations. A number of factors have contributed to this relative failure. Partly, it is due to the difficulty in starting up completely new manufacturing lines employing novel technologies. The well-established silicon industry has a strong manufacturing infrastructure based on the global semiconductor industry and is supported by an excellent knowledge base and extensive characterisation techniques. In contrast, the thin-film technologies were new and production lines were one-offs, requiring major developments of process equipment, often at much higher initial capital costs than for silicon. With both the process equipment itself and the upscaling of the thin-film technology having to be produced at the same time, long delays were experienced in achieving good yields and product costs were not as low as planned. Crucially, silicon technology was able to continuously improve solar cell efficiency and reduce manufacturing costs. In the 1980s, it was widely believed that a minimum solar cell efficiency of 10% was essential to achieving competitive costs for photovoltaic-generated electricity based on the area-related costs for BOS components [184]. Silicon modules were already above that efficiency, but the different thin-film technologies did not achieve it until the early 2000s – by which time crystalline silicon modules were approaching 20%. The defining period was 2003 onward, when the market was expanding rapidly and decisions needed to be taken about investments in production capacity at the billion-dollar level. The choice was between proven silicon-based technology with a strong infrastructure and guaranteed production equipment performance backed by a good record of in-service performance and unproven technologies of uncertain production costs. The decision globally was overwhelmingly in favour of silicon. It was only the undersupply of silicon product to the booming market that gave an opportunity to thin films, which was quickly taken by First Solar – but the market share then declined as the silicon supply chain was debottlenecked. Crystalline silicon remains the dominant photovoltaic technology and is expected to do so for some time, with continuing improvement in module efficiency, longevity, and costs, as described in the next chapter.

References

1 P. Rappaport: Proceedings of PV Conversion of Solar Energy for Terrestrial Applications NSF-RA-N-74-013, Cherry Hill (1974) 1–10.
2 S.N. Dermatis and J.W. Faust: IEEE Transactions in Communications and Electronics 81 (1983) 94–98.
3 R.G. Siedensticker et al.: Proceedings of the 11th IEEE PVSC (1975) 299–302.
4 S. Narashima et al.: Solid State Electronics 42 (1998) 1631–1640.
5 W. Koch et al. in 'Handbook of PV Science and Engineering' eds A. Luque and S. Hegedus pub: Wiley (2003) 239.
6 Pittsburg Business News 7 November 2002.

7 K.V. Ravi et al.: Proceedings of the 11th IEEE PVSC (1975) 280–289.

8 M.J. Kardauskas et al.: Proceedings of the 25th IEEE PVSC (1996) 383–388.

9 W. Schmidt et al.: Progress in Photovoltaics Science and Applications 10 (2002) 129–140.

10 G. Hahn and P. Geiger: Progress in Photovoltaics Science and Applications 11 (2003) 341–346.

11 RWE Press Release 22 October 2002.

12 T.M. Bruton et al.: Final Publishable Report APAS MUSIC FM Project (1996) European Commission.

13 PV Magazine 11 December 2012.

14 E. Sachs: PhD thesis Massachusetts Institute of Technology 27 April 1983.

15 E. Sachs and D Ely: Proceedings of the 17th IEEE PVSC (1984) 1418–1419.

16 J.I. Hanoka: Proceedings of the 29th IEEE PVSC (2002) 66–69.

17 G. Hahn and P. Geiger: Progress in Photovoltaics Science and Applications 11 (2003) 341–346.

18 https://www.nytimes.com/2011/01/15/business/energy-environment/15solar.html accessed 7 August 2020.

19 www. ventizz.de/en/pm100324php. accessed 13 July 2018.

20 Y. Maeda et al.: Proceedings of the 16th IEEE PVSC (1982) 133.

21 Y. Maeda et al.: Proceedings of the 17th IEEE PVSC (1984) 1112–1115.

22 M. Suzuki et al.: Journal of Crystal Growth 104 (1990) 102–107.

23 Y. Hatanaka: Proceedings of the 22nd IEEE PVSC (1991) 859–863.

24 T. Saitoh et al.: Progress in Photovoltaics Science and Applications 1 (1993) 11–23.

25 H. Lange et al.: Journal of Crystal Growth 104 (1990) 108.

26 A. Burgers et al.: Proceedings of the 21st EUPVSEC (2006) 651–654.

27 A. Schoenecker et al.: Proceedings of the 26th EUPVSEC (2011) 920–924.

28 Y.P. Botchak Mouafi et al.: Solar Energy and Materials 146 (2016) 25–34.

29 http://www.rgsdevelopment.nl/ accessed 7 August 2020.

30 Arpa-e Open 2009 Program 'Direct Wafer enabling Terrawatt Photovoltaics'.

31 US patent 8293009 granted 23 October 2012.

32 E. Sachs et al.: Proceedings of the 28th EUPVSEC (2013) 907–910.

33 https://1366tech.com/about-1366/ accessed 7 August 2020.

34 H. Tayanaka et al.: Proceedings of the of the 6th Sony Research Forum (1996) 556 [in Japanese].

35 R. Brendel: Proceedings of the 14th EUPVSEC (1997) 1354–1357.

36 R. Bergmann and T.J. Rinke: Progress in Photovoltaics Science and Applications 8 (2000) 451–456.

37 J.H. Petermann et al.: Progress in Photovoltaics Science and Applications 20 (2012) 1–5.

38 S. Janz et al.: Proceedings of the 31st EUPVSEC (2015) 288–291.

39 https://www.nexwafe.com/ accessed 7 August 2020.

40 E.A. Alsema: Progress in Photovoltaics Science and Applications 8 (2000) 17–25.

41 S. Wagner: Proceedings of the 16th IEEE PVSC (1982) 685–691.

42 D.C. Reynolds: Physics Review 96 (1954) 533–534.

43 L.D. Massie: Proceedings of the 5th IEEE PVSC (1965) C-1-1–C-1-13.

44 F.A. Shirland and J.R. Heitenan: Proceedings of the 5th IEEE PVSC (1965) C-3-1–C-3-13.

45 J. Perlin: 'From Space to Earth, the story of solar electricity' pub: Harvard University Press (2000) 173.

46 K.W. Boer in 'Solar Power for the World' ed. W. Palz pub: Pan Stanford (2014) 220.

47 R.B. Hall et al.: Proceedings of the 15th IEEE PVSC (1981) 777–779.

48 G.H. Herwig et al.: Proceedings of the 16th IEEE PVSC (1982) 713–718.

49 K.W. Boer in 'Solar Power for the World' ed. W. Palz pub: Pan Stanford (2014) 222.

50 R. Crandall and W. Luft: Progress in Photovoltaics Science and Applications 3 (1995) 315–331.

51 D.E. Carlson and C. Wronski: Applied Physics Letters 28 (1976) 671.

52 D.E. Carlson: Proceedings of the 14th IEEE PVSC (1980) 291–297.

53 A. Madan et al.: Applied Physics Letters 37 (1980) 826.

54 Y. Tawada et al.: Applied Physics Letters 39 (1981) 237.

55 A. Catalano et al.: Proceedings of the 16th IEEE PVSC (1982) 1421–1422.

56 D.L. Staebler and C. Wronski: Applied Physics Letters 21 (1977) 292.

57 M.A. Green et al.: Progress in Photovoltaics Science and Applications 25 (2017) 3–13.

58 S. Hegedus et al.: Proceedings of the 25th IEEE PVSC (1996) 1129–1133.

59 K. Winz et al.: Proceedings of the 25th IEEE PVSC (1996) 1149–1152.

60 R.R. Arya and D.E. Carlson: Progress in Photovoltaics Science and Applications 10 (2002) 69–76.

61 J. Yang et al.: Proceedings of the 20th IEEE PVSC (1988) 241.

62 R.E.I. Schropp in 'Solar Cell Materials – Developing Technologies' eds G.J. Conibeer and A. Willoughby pub: Wiley (2014) 103.

63 J. Meier et al.: Materials Research Society Symposium Proceedings 420 (1996) 3.

64 E. Moulin et al. in 'PV Solar Energy – From Fundamentals to Applications' eds A. Reinders et al. pub: Wiley (2014) 218.

65 K. Nishimura et al.: Proceedings of the PVSEC-1 (1984) P-1-15.

66 http://www.kaneka.be/new-business accessed 7 August 2020.

67 G. Janes and L. Bouamane: 'Power from Sunshine – A Business History of Solar Energy' pub: Harvard Business School (2012) 12–105.

68 B. Mc Nelis in 'Clean Energy from Photovolatics' eds M.D. Archer and R. Hill pub: World Scientific (2001) 724.

69 Bekaert Financial Report – Press Release (2002).

70 https://en.wikipedia.org/wiki/Energy_Conversion_Devices accessed 7 August 2020.

71 A. Delahoy: Subcontract Report SERI/STR-211-205 (August 1983).

72 D.E. Carlson and S. Wagner in 'Renewable Energy Sources for Fuel and Electricity' eds T.B. Johansson and L. Barnham pub: Island Press (1993) 422.

73 J.P.M. Schmitt: Thin Solid Films 174 (1989) 193–202.

74 The Washington Post 27 May 1994.

75 P. Varadi: 'Sun Above the Horizon: The Meteoric Rise of the Solar Industry' pub: Pan Stanford Publishing (2014) 229–231.

76 https://en.wikipedia.org/wiki/TEL_Solar accessed 7 August 2020.

77 J. Cashmore et al.: Progress in Photovoltaics Science and Applications 23 (2015) 1441–1447.

78 J.J. Loferski et al.: Proceedings of the 14th IEEE PVSC (1980) 375–380.

79 A. Goertzberger: Proceedings of the 15th IEEE PVSC (1981) 867–870.

80 S.B. Schuldt et al.: Proceedings of the 15th IEEE PVSC (1981) 934–940.

81 A.M. Barnett: Proceedings of the 18th IEEE PVSC (1985) 1094–1099.

82 A.M. Barnett: Proceedings of the 25th IEEE PVSC (1996) 1–8.

83 A.M. Barnett et al.: Progress in Photovoltaics Science and Applications 2 (1994) 163–170.

84 A.M. Barnett: Proceedings of the 14th EUPVSEC (1997) 999–1002.

85 B. Sopori in 'Handbook of Photovoltaic Science and Engineering' eds A. Luque and S. Hegedus pub: Wiley (2003) 318.

86 J. Culik et al.: Proceedings of the 29th IEEE PVSC (2002) 392–394.

87 Renewable Energy World 22 February 2004.

88 Z. Shi and S. Wenham: Progress in Photovoltaics Science and Applications 2 (1994) 153–162.

89 T.M. Bruton et al.: Physica Status Solidi 154 (1996) 623–633.

90 O. Gabriel et al. in 'PV Solar Energy – From Fundamentals to Applications' eds A. Reinders, P. Verlinden, W. van Sark, and A. Freundlich pub: Wiley (2014) 226–237.

91 S.A. Edmiston: Progress in Photovoltaics Science and Applications 3 (1995) 333–350.

92 M.J. Keevers et al.: Proceedings of the 22nd EUPVSEC (2007) 1783–1790.

93 R.W. Stenhagen and R Wueber: 'The Handbook of Research on Energy and Entrepreneurship' pub: Edward Elgar (2011) 98.

94 P.A. Basore: Proceedings of the 29th IEEE PVSC (2002) 392–394.

95 M.J. Keevers et al.: Proceedings of the 23rd EUPVSEC (2007) 1783–1789.

96 R. Egan et al: Proceedings of the 24th EUPVSEC (2009) 2279–2285.

97 E. Wesoff: Green Tech Media 1 December 2015.

98 J.L. Shay and S. Wagner: Applied Physics Letters 27 (1975) 89.

99 L.L. Kazmerski et al.: Applied Physics Letters 29 (1976) 268.

100 M. Spitzer et al.: Proceedings of the 14th IEEE PVSC (1980) 585–590.

101 R.A. Mickelson and W.S. Chen: Proceedings of the 15th IEEE PVSC (1981) 800–804.

102 H.W. Schock and R. Noufi: Progress in Photovoltaics Science and Applications 8 (2000) 151–160.

103 J. Pietoszewski et al.: Proceedings of the 14th IEEE PVSC (1980) 980–985.

104 V.J. Kapur et al.: Proceedings of the 18th IEEE PVSC (1985) 1429–1432.

105 J.H. Ermer et al.: Proceedings of the 18th IEEE PVSC (1985) 1655–1658.

106 R.A. Mickelson et al.: Proceedings of the 18th IEEE PVSC (1985) 1069–1073.

107 W. Giriat and J. Stankiewicz: Proceedings of the 14th IEEE PVSC (1980) 800–804.

108 R.C. Powell et al.: Proceedings of the 18th IEEE PVSC (1985) 1050–1053.

109 J.J. Loferski et al.: Proceedings of the 15th IEEE PVSC (1981) 1056–1061.

110 W. Arndt et al.: Proceedings of the 18th IEEE PVSC (1985) 1671–1676.

111 K. Zweibel et al.: Proceedings of the 18th IEEE PVSC (1985) 1393–1398.

112 J.R. Tuttle et al.: Proceedings of the 23rd IEEE PVSC (1993) 415–421.

113 A.M. Gabor et al.: Applied Physics Letters 65 (1994) 198–200.

114 J. Hedstrom et al.: Proceedings of the 23rd IEEE PVSC (1993) 364–371.

115 H.W. Schock and R. Noufi: Progress in Photovoltaics Science and Applications 8 (2000) 151–160.

116 I. Repins et al.: Progress in Photovoltaics Science and Applications 16 (2008) 235–239.

117 P. Jackson et al.: Progress in Photovoltaics Science and Applications 19 (2011) 894–897.

118 https://www.zsw-bw.de/en/newsroom/news/news-detail/news/detail/News/zsw-stellt-neuen-weltrekord-bei-duennschicht-solarzellen-auf.html accessed 7 August 2020.

119 http://www.solar-frontier.com/eng/news/2019/0117_press.html accessed 7 August 2020.

120 M.A. Green et al.: Progress in Photovoltaics Science and Applications 17 (2009) 347–349.

121 S. Yang et al.: Progress in Photovoltaics Science and Applications 23 (2015) 1157–1163.

122 T. Fuerer et al.: Progress in Photovoltaics Science and Applications 25 (2017) 645–667.

123 P.L. Morel et al.: Proceedings of the 18th IEEE PVSC (1985) 876–882.

124 C. Eberspacher et al.: Proceedings of the 18th IEEE PVSC (1985) 1031–1035.

125 C. Frederic et al.: Proceedings of the 23rd IEEE PVSC (1993) 437–440.

126 K. Zweibel et al.: Proceedings of the 25th IEEE PVSC (1996) 745–750.

127 T. Dalibor et al.: Proceedings of the 25th EUPVSEC (2010) 2854–2857.

128 https://www.avancis.de/en/company/ accessed 7 August 2020.

129 M. Powalla et al.: Proceedings of the 29th IEEE PVSC (2002) 571–574.

130 https://www.manz.com/en/markets/solar/cigs-fab/ accessed 7 August 2020.

131 https://en.wikipedia.org/wiki/Nanosolar accessed 7 August 2020.

132 PV Magazine 10 January 2020.

133 M.A. Green et al.: Progress in Photovoltaics Science and Applications 28 (2020) 3–15.

134 K. Kushiya et al.: Japanese Journal of Applied Physics 33 (Part 1) 12A (1994) 285.

135 K. Kushiya et al.: Proceedings of the 29th IEEE PVSC (2002) 579–582.

136 D.A. Cusano: Solid State Electronics 6 (1963) 217.

137 L.D. Massie: Proceedings of the 5th IEEE PVSC (1965) C-1-1–C-1-13.

138 P.V. Meyers and R.W. Birkmire: Progress in Photovoltaics Science and Applications 3 (1995) 393–402.

139 D. Bonnet and H. Rabenhorst: Proceedings of the 9th IEEE PVSC (1972) 129.

140 Y.-S. Tyan and E.A. Perez-Albuene: Proceedings of the 16th IEEE PVSC (1982) 794–800.

141 B.M. Basol et al.: Proceedings of the 16th IEEE PVSC (1982) 805–808.

142 H. Uda et al.: Proceedings of the 16th IEEE PVSC (1982) 801–804.

143 H.B. Sereze: Proceedings of the 15th IEEE PVSC (1981) 1068–1072.

144 C. Ferekides et al.: Proceedings of the 23rd IEEE PVSC (1993) 389–393.

145 J.M. Woodcock et al.: Proceedings of the 22nd IEEE PVSC (1991) 842–847.

146 P.V. Meyers and S.P. Albright: Progress in Photovoltaics Science and Applications 8 (2000) 161–169.

147 M.A. Green et al.: Progress in Photovoltaics Science and Applications 21 (2013) 1–11.

148 M.A. Green et al.: Progress in Photovoltaics Science and Applications 26 (2018) 3–12.

149 https://www.japantimes.co.jp/news/2008/11/19/business/panasonics-takeover-of-sanyo-all-about-green-technology/ accessed 7 August 2020.

150 D.W. Cunningham et al.: Proceedings of the 29th IEEE PVSC (2002) 559–562.

151 D. Bonnet et al.: Proceedings of the 29th IEEE PVSC (2002) 563–566.

152 D. Bonnet in 'Solar Power for the World' ed. W. Palz pub: Pan Stanford (2014) 635–641.
153 J.F. Nolan: Proceedings of the 23rd IEEE PVSC (1993) 34–41.
154 B.E. McCandless et al.: Proceedings of the 29rd IEEE PVSC (2002) 547–550.
155 D. Rose et al.: Proceedings of the 29rd IEEE PVSC (2002) 555–559.
156 https://cleantechnica.com/2014/03/26/average-manufacturing-costs-solar-halve-gas-prices/ accessed 7 August 2020.
157 V.M. Fthenakis and P.D. Moscowitz: Progress in Photovoltaics Science and Applications 8 (2000) 27–38.
158 First Solar press release 26 April 2018.
159 J. Sites et al.: Proceedings of the 33rd EUPVSEC (2017) 998–1000.
160 W. Palz in 'Solar Power for the World' ed. W. Palz pub: Pan Stanford (2014) 38.
161 K. Hara and H. Arakowa in 'Handbook of Photovoltaic Science and Engineering' eds A. Luque and S. Hegedus pub: Wiley (2003) 663–700.
162 H. Gerischer and H. Tributsch: Berichte der Bunsengesellschaft für physikalische Chemie 72 (1968) 437–445.
163 B. O'Regan and M. Gratzel: Nature 353 (1991) 737–740.
164 M. Gratzel: Progress in Photovoltaics Science and Applications 8 (2000) 171–185.
165 B. Kippelen: 'PV Solar Energy – From Fundamentals to Applications' eds A. Reinders, P. Verlinden, W. van Sark, and A. Freundlich pub: Wiley (2017) 259.
166 J.H. Burroughes et al.: Nature 347 (1990) 539–541.
167 B.P. Rand et al.: Progress in Photovoltaics Science and Applications 15 (2007) 659–676.
168 N.C. Greenham: Philosophical Transactions of the Royal Society A 371 (2013) 2011.0414.
169 J. Brabec et al.: Advanced Materials 22 (2010) 3839–3856.
170 M.A. Green et al.: Progress in Photovoltaics Science and Applications 28 (2020) 3–15.
171 S.D. Stranks and H.J. Snaith in 'PV Solar Energy – From Fundamentals to Applications' eds A. Reinders, P. Verlinden, W. van Sark, and A. Freundlich pub: Wiley (2017) 275–291.
172 H.S. Kim et al.: Scientific Reports 2 (2012) 591.
173 M.M. Lee et al.: Science 338 (2012) 643–647.
174 L. Ba et al.: Progress in Photovoltaics Science and Applications 26 (2018) 924–933.
175 K.A. Bush et al.: Nature Energy 2(4) (2017) 17009.
176 R.M. Swanson: "Handbook of Photovoltaic Science and Engineering" ed A. Luque and S. Hegeduspub J. Wiley (2003) 449–504.
177 M.W. Edenburn and E. Boes: : Proc 17th IEEE PVSC (1984) 473–484.
178 T.M. Bruton: Proceedings of the 16th German National PV Symposium (2002).
179 C. Jennings et al.: Proceedings of the 25th IEEE PVSC (1996) 1513–1516.
180 V. Garboushian: Proceedings of the 25th IEEE PVSC (1996) 1373–1376.
181 APS Presentation: International Conference on Solar Concentration for the Generation of Electricity or Hydrogen May 2005, Scottsdale, AZ, USA.
182 H. Lerchenmuller et al.: 4th International Conference on Solar Concentration for the Generation of Electricity or Hydrogen (2007) 225–228.
183 https://www.soitec.com/en/company/soitec-in-brief/history accessed 7 August 2020.
184 H.W. Brandhorst Jr: Proceedings of the 17th IEEE PVSC (1984) 1–6.

8

Current Status of Crystalline Silicon Manufacturing and Future Trends

8.1 Introduction

Chapter 7 described how a wide range of alternative technologies have been explored in multidecade and multibillion-dollar research programmes. Despite the potential advantages of thin-film technologies, crystalline silicon remains the dominant production technology, having supplied at least 90% of the 118 GWp photovoltaics installed in 2018 [1,2], and it is expected to continue to be so for the foreseeable future. This chapter aims to describe the current status of silicon manufacturing and to highlight the future directions in which the silicon wafer-based technology might be expected to develop.

Chapter 4 described the development from the earliest small-scale, batch-processed silicon solar cells to gigawatt-scale, highly automated in-line manufacturing plants. This was based on quite similar technology for both monocrystalline and multicrystalline silicon wafers. Both were p type wafers of typically 1 Ω/cm resistivity. Both had textured surfaces, a SiN_x ARC, an aluminium BSF across the whole cell rear, and screen-printed contacts with silver on the front side and a silver aluminium gridded contact on the rear. In 2011, typical crystalline silicon modules had efficiencies in the range 12–16%, with premium modules in the range 18–20% [3]. Efficiencies are now much higher, with standard modules in the range 16–19% and premium modules >20% [4]. This chapter thus also looks at how the production technology has improved with novel cell designs and improved materials, as well as new concepts based on passivated contacts, the use of n type wafers, and rear heterojunction cells, which have combined to achieve the current world record for silicon solar cell efficiency at 26.6% [5]. It further discusses the routes to achieving even higher efficiencies for tandem cells based on silicon in order to bring about the terawatt markets predicted for the future.

Photovoltaics from Milliwatts to Gigawatts: Understanding Market and Technology Drivers toward Terawatts,
First Edition. Tim Bruton.
© 2021 John Wiley & Sons Ltd. Published 2021 by John Wiley & Sons Ltd.

8.2   Approaches to High-Efficiency Silicon Solar Cells on p Type Silicon Wafers

The upper limit for a single-junction silicon solar cell under AM1.5 insolation has long been accepted as 29% [6]. Swanson indicated that achieving this level beyond the then 25% best was dependent on very effective passivation [6]. Glunz has described the process of attaining such very high efficiencies as being like successively plugging multiple holes in a leaky bucket [7]. The principal actions that must be taken are listed in Table 8.1. It is pursing these actions, both singly and in concert, that has led to the continued improvement in both laboratory and production solar cells.

8.2.1   LGBC Solar Cells

While the dominant process used in the manufacture of silicon solar cells has traditionally been screen printing, this comes with some well-known disadvantages, especially for the front contact. The emitter must be relatively heavily doped, both to allow good ohmic contact with the silver paste and to prevent shunting of the junction, which limits the blue response of the solar cell [3]. An early attempt to overcome this was the laser-grooved buried-contact (LGBC) cell, developed at the University of New South Wales in the early 1980s (Figure 8.1) [8]. The key feature is that the front metal contact is not screen-printed, but plated into a groove in the silicon surface created by laser scribing.

The manufacturing sequence has been described for the pilot line established by BP Solar in Madrid in 1989 [9]. A standard $1\,\Omega/cm$ p type CZ wafer is textured and lightly diffused, and LPCVD silicon nitride ARC is applied. The laser grooves and bus bar are cut through the ARC, forming grooves about $20\,\mu m$ wide and $35\,\mu m$ deep. The grooves are then etched to remove the amorphous silicon phase deposited in them. Clean crystallographic facets are revealed and the groove dimensions are increased by $10\,\mu m$.

Table 8.1   Main causes of loss in silicon solar cells

Area for improvement	Actions
Metallisation	Better pastes/plating
Front side	Selective emitters
Rear	Dielectric passivation
Base	Transition to n type
Optical losses	Rear-contact cells
Last recombination	Passivated contacts

Source: S.W. Glunz: Proceedings - EU PVSEC 2012, 27th European Photovoltaic Solar Energy Conference and Exhibition, 2BP.12. © 2012, WIP GmbH & Co Planungs-KG. [7]

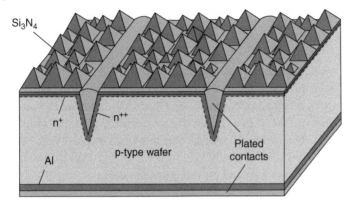

Figure 8.1 Structure of an LGBC solar cell (*Source:* Courtesy BP Archive.)

The grooves are then heavily diffused to around $10\,\Omega$ per square. Aluminium is evaporated on to the cell rear and driven in to form a back surface field (BSF). The cell is plated using an electroless nickel/copper/silver deposition [10]. Edge isolation is effected by laser scribing and cleaving off of the cell edges.

Cell efficiencies up to 18% have been achieved in the initial start-up phase on 100 mm p square Czochralski (CZ) wafers. One of the earliest applications was the installation of 550 kWp at the Union Fenosa plant near Toledo, Spain, which performed very well until being decommissioned in 2014 [11]. The wafer size was increased to 125 mm by 1994 [12], and by 2006 large-area cells ($147.4\,cm^2$) were demonstrating 20.1% efficiency using laser-fired rear contacts on a float zone (FZ) wafer [13].

One of the initial advantages of the LGBC approach was that the high volume of highly conducting copper in the groove coupled with the highly doped emitter made for a very low series resistance, making the cell useful in concentrated sunlight applications. Cells could be used at 5 times 1 sun insolation (5×) without loss of efficiency [14]. Additional grid lines could be scribed at minimal additional cost and without significant surface shading. An early result with prismatic covers designed to deflect light away from the grid lines gave 20% efficiency at 20× for use in linear Fresnel lens concentrating systems [15]. Further work showed that an efficiency of 18.7% at 100× on a small-area cell ($2.6\,cm^2$) was achievable [16]. These cells were developed for use at 120x in a point-focus system with the potential to achieve a €1.6/Wp AC system cost – well below the flat plate cost at the time [17]. However, as described in Chapter 7, flat plate costs fell so dramatically that these low-concentration approaches became uncompetitive.

While most of the development and production was performed on monocrystalline material, it was shown that good efficiencies could be obtained on multicrystalline silicon, too. A screen-printed aluminium was used to produce the BSF. It was possible to produce a $144\,cm^2$ cell with an efficiency of 16.9%, which was amongst the highest values for a large-area multicrystalline silicon solar cell at the time [18].

In total, over 120 MWp of LGBG cells were produced by BP Solar, but production was terminated in 2009 as part of BP's withdrawal from the sector.

A variant of the technology produced by the Suntech Company in China achieved over 20% efficiency, but manufacturing was again terminated due to problems with cell adhesion [19] and the company later filed for bankruptcy.

Although the LGBG technology is no longer in production, it successfully demonstrated the use of selective emitters and plating technology, both of which are now undergoing further development, and illustrated the system cost benefits of high solar cell efficiency.

8.2.2 Selective Emitters

The advantages of selective emitters in improving ohmic contact to the metallisation and enhancing the blue response of solar cells are well known [20]. A variety of techniques have been used to produce the localised highly doped region. One of the first involved performing a relatively heavy diffusion, masking the grid-line contact area, and etching back the emitter to a typical $140\,\Omega$ [21]. However, this process was difficult to control, so the preferred method is to perform a light surface diffusion followed by a second heavy diffusion in the contact region, as discussed for LGBC cells.

The method used for the second diffusion varies. In its simplest form, a conventional surface diffusion is carried out and the phosphorus silicate glass is left in place and locally melted with a laser to form the heavily doped region, decreasing the contact resistance by a factor of 100 [22]. Direct methods can also be used, such as applying the silicon nitride ARC and then using a laser chemical process, in which a laser beam is steered down a jet of phosphoric acid, etching through the ARC and creating the deep junction [23]. However, these approaches create a difficulty in precisely applying the screen-print metallisation to the heavily doped region, and high-resolution printing is required [24].

In general, the efficiency gains from a selective emitter are between 0.5 and 1.0% absolute, which do not justify the increased complexity of the manufacturing process [20]. However, the selective-emitter approach can be used as part of a high-efficiency manufacturing process, such as for passivated emitter and rear contact (PERC) cells [25].

8.2.3 PERL and PERC Solar Cells

In the early 1980s, the volume of manufacture of silicon solar cells was based on a relatively simple cell design of heavily doped emitter, screen-printed contacts on front and rear, and no BSF or surface passivation. A 12% solar cell efficiency was typical for these devices on monocrystalline silicon, and around 10% on multicrystalline. It was recognised at that time that an improvement in efficiency was essential to achieve lower costs. In the United States, a research target of a 20% solar cell efficiency by 1986 was tabled and Spire Corporation achieved an 18% cell by applying a nonpassivating ARC and using phosphorus ion implantation to produce an emitter free from precipitates [26]. It was quickly realised that passivation of the front surface was

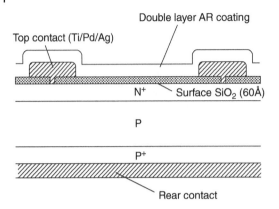

Figure 8.2 Structure of a PESC [27] (*Source:* M.A. Green et al: Proceedings of the 18th IEEE Photovoltaic Specialists Conference, 39-42. © 1985, IEEE. Courtesy IEEE.)

necessary to achieve a 20% efficiency, and with good surface passivation a 20.9%-efficiency passivated emitter solar cell (PESC) was reported on a 0.2 Ω/cm FZ wafer by the UNSW group in 1985 [27]. In this cell, a passivation layer of silicon dioxide 60 Å thick was used. Narrow vias in the oxide were made for the Ti/Pd/Ag contact, and a double-layer ARC was utilised. The rear surface was planar with a BSF and full metal coverage. All of this is shown in Figure 8.2.

Reflection from the front surface was minimised by using micro-grooves formed through photolithography, which lowered reflection losses to 1% – significantly better than the standard random-textured surface. The rear surface remained a significant source of recombination, as demonstrated in later modelling (see Figure 8.3). The solar cell efficiency was modelled against wafer thickness for different rear surface recombination velocities and wafer lifetimes [28]. For the CZ wafers, a minority carrier diffusion length of 150 μm was assumed, while 1200 μm was assumed for FZ silicon. Also assumed were a constant high rear surface reflectivity of 95% and good front surface passivation. The highest rear surface recombination velocity of 1×10^6 cm/s shown in Figure 8.3 corresponds to an unpassivated surface, that of 2000 cm/s corresponds to a typical screen-printed aluminium BSF, while that of 250 cm/s corresponds to a well-passivated surface. The upper curve shows the results of a high-lifetime FZ wafer. These results demonstrate that 20% cell efficiency can only be achieved for a standard CZ material if good rear surface passivation is applied.

The next stage in the development of the high-efficiency concept was thus to apply a passivated rear contact, giving rise to the PERC cell (reported as achieving 22.8% efficiency). This had a number of new features as compared to the previous PESC. The rear surface was passivated with an oxide, similar to the front side, and vias were opened for the metal contact, which covered the passivating oxide, providing good rear surface reflection. The microgrooves were replaced by inverted pyramids formed through photolithography [29], reducing surface reflection compared to the micro-groove case. Further efficiency improvement came from a heavy p+ diffusion in the rear contact, forming the passivated emitter-rear locally diffused (PERL) cell, but its simpler processing (still with very high efficiency) made the PERC cell the vehicle chosen for large-scale manufacture [20].

The structure of the PERL cell is illustrated in Figure 8.4. This was first reported in 1990, with very low reflection losses and an internal quantum efficiency close to 100%

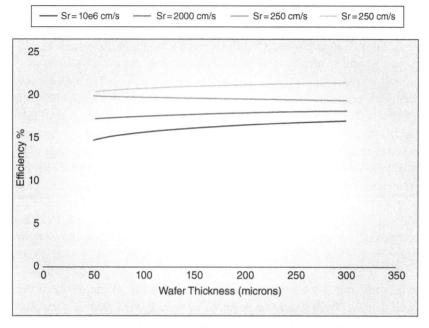

Figure 8.3 Modelled efficiency versus wafer thickness for varying degrees of rear surface passivation, assuming a CZ minority carrier diffusion length of 150 μm [28] *Source:* T.M. Bruton et al: Proceedings - EU PVSEC 2001, 17th European Photovoltaic Solar Energy Conference and Exhibition, 1282-1286. © 2001, WIP GmbH & Co Planungs-KG.

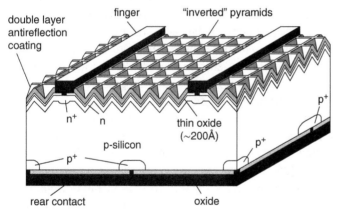

Figure 8.4 Structure of a PERL silicon solar cell [32] (*Source:* J. Zhao et al: Progress in Photovoltaics-Research and Applications, 7, 471-47. © 1999, John Wiley & Sons.)

up to wavelengths of 1.1 μm [30]. It was later further improved to give a 23.1% (4 cm^2) cell efficiency, independently certified with the AM1.5 global spectrum. A 2 Ω/cm p type FZ wafer was processed with 1100 Å of front oxide on the inverted pyramids (to give a good ARC) and 3000 Å of oxide on the rear surface [31]. The fill factor was relatively low at 81.0%. Some of the fill factor loss was attributed to current crowding around the localised rear contacts, which was overcome by p type doping across the whole rear surface (while maintaining a highly doped p+ region in the contacts), producing the passivated emitter totally rear-diffused (PERT) cell. A $4\,cm^2$ PERT cell with a ZnS/MgF_2 double-layer ARC on a $4.8\,\Omega/cm$ MCZ substrate gave a 24.5% efficiency, but a PERL cell with the same process on a $1.0\,\Omega/cm$ FZ wafer achieved 24.7% [32]. This remained the world-record silicon solar cell efficiency for the next 15 years [33].

While the PERL structure has produced very high solar cell efficiencies, it necessitates a complex manufacturing process involving up to six photolithographic steps [34], and commercial preference is thus given to the simpler PERC structure, which brings only a moderate efficiency loss [20]. There was a long-term effort to make the PERC structure cost-effective to manufacture by 2008, with many laboratories around the world contributing to industrialising the process. There were two main approaches: increasing solar cell efficiency and moving to thinner silicon wafers. The latter became particularly important in the period 2000–2010, when the industry was growing rapidly and thus outstripping the supply of polysilicon feedstock, leading to spot prices of up to €300/kg (as compared to a contract price of €30/kg). It was anticipated that wafer thicknesses well below $200\,\mu m$ would be required to reduce costs [35]. A particular problem was identified with the aluminium BSF approach, where bowing of the wafer could occur on firing of the thick aluminium paste on the cell rear. New approaches to passivating the rear surface were thus needed for very thin wafers [36]. The PERC cell structure met this need. However, after 2010, the emphasis changed as the polysilicon supply issues eased and wafer thickness stabilised at around $180\,\mu m$. There were now a large number of manufacturers at the gigawatt scale offering similar products at low cost, and the decline in module price made the balance-of-systems cost more significant to the final system price. Therefore, manufacturers sought to distinguish themselves from the competition by having a higher-efficiency product with low installation costs. The principal challenges were to remove the photolithographic steps and to introduce established in-line high-throughput processes.

One of the first steps toward simplifying the process was the use of the conventional manufacturing process for random texturing of the silicon surface, replacing the inverted pyramids. This resulted in small-area cells ($4\,cm^2$) of 21.0% efficiency with only two photolithographic steps [35]. Next was the move from using thermal silicon oxide passivation, which is a slow batch process, to the well-established high-throughput industrial process for PECVD silicon nitride. This gave a good Voc in initial trials, but a low efficiency of 17.8% on a nontextured silicon solar cell [37]. Forming the rear contact was also a challenge without photolithography. FhG.ISE demonstrated that it was possible to use lasers to rapidly form point contacts on a $2\,\mu m$ aluminium layer over the rear dielectric passivation. Efficiencies over 20% were achieved with silicon dioxide and over 19% with the industrially compatible PECVD silicon nitride passivation [35]. The front metallisation also needed amending from the evaporated Ti/Pd/Ag contacts of the laboratory cells. Firing through a silver paste on PECVD SiN_x was well established. It was demonstrated that this could be used in the PERC process [36], and good efficiencies were obtained on monocrystalline silicon. However, the use of screen-printed aluminium over the silicon nitride rear passivating layer actually increased rear surface recombination compared to the full BSF, so an alternative PECVD silicon oxide/silicon nitride stack was used instead, increasing the efficiency. A full 100 cm^2 multicrystalline silicon solar cell gave 17.4% efficiency: about 1% absolute higher than the equivalent Al-BSF cell [38]. This work confirmed the importance of the rear dielectric acting to reflect the longer-wavelength light in the PERC structure, as shown in Figure 8.5.

Laser ablation was used to open vias in order to achieve 16% efficiency on a printed rear aluminium contact, as an alternative to laser firing [39]. The use of the silicon oxide/nitride stack was confirmed by other groups, showing a 17.4% PERC cell (147.4 cm^2) with screen-printed contacts [40].

Further progress in identifying a thermally stable rear-surface passivation was achieved by the application of atomic layer deposition for aluminium oxide (ALD-Al$_2$O$_3$). This passivation matched PERC cells with good thermal oxide rear-surface passivation, and a 30 nm ALD-Al$_2$O$_3$ passivated cell with a 200 nm PECVD SiO$_x$ capping layer gave 20.6% efficiency – better than the thermal oxide control cell [41].

All the stepping stones were now in place for the next stage in the industrialisation of the PERC process. This was the announcement in 2011 by Q-Cells that its Q.ANTUM high-efficiency solar cell was entering pilot production [42]. A pilot line running at 100–500 cells per week was used to process both monocrystalline and multicrystalline solar cells. The device structure is shown in Figure 8.6. A PECVD silicon nitride front-side passivation was used with fine-line screen printing thickened by plating [43]. The rear-side dielectric was not specified, and laser firing was used to form the rear contacts. A world-record 19.5% efficiency was verified for a large-area multicrystalline solar cell (243 cm^2), with 20.3% efficiency for a CZ monocrystalline silicon cell. The front side was textured, while the rear surface was relatively smooth.

Figure 8.5 Total reflectance from an aluminium-coated silicon oxide/nitride stack in a PERC cell [38] (*Source:* P. Choulat et al: Proceedings - EU PVSEC 2007, 22nd European Photovoltaic Solar Energy Conference and Exhibition, 1011-1014. © 2007, WIP GmbH & Co Planungs-KG. Courtesy EUPVSEC.)

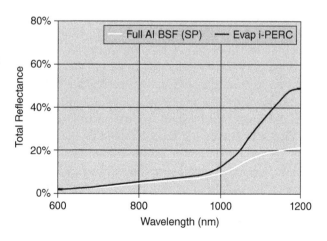

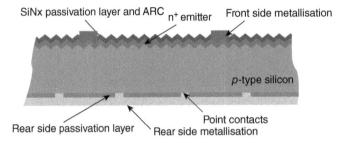

Figure 8.6   Q.ANTUM PERC solar cell [42] (*Source:* P. Engelhart et al: Proceedings - EU PVSEC 2011, 26th European Photovoltaic Solar Energy Conference and Exhibition, 821-826. © 2011, WIP GmbH & Co Planungs-KG. Courtesy EUPVSEC.)

Further impetus for the large-scale production of the PERC cell was given by the process equipment industry delivering turnkey production lines [44]. A wet etching tool was modified from the standard process line to allow edge isolation of the surface diffusion and smoothing of the rear surface, while keeping the surface texture intact without masking. A high-throughput machine was used for deposition of the ALD-Al_2O_3, while PECVD silicon nitride was used for the front passivation and for simultaneous capping of the ALD-Al_2O_3, which could be as thin as 4 nm. Laser ablation was used to open the vias in the rear dielectric and aluminium screen printing was used to form the rear contact. The correct combination of laser and aluminium paste removed any requirement for annealing laser damage after opening the vias. A net gain of 1% absolute efficiency was found for the PERC process over the Al-BSF for multicrystalline silicon, with the best cell batches averaging 18% efficiency. Using PECVD Al_2O_3/SiN_x, it was possible to achieve 20.3% efficiency for PERC monocrystalline silicon cells in an industrial line [45]. Trina Solar reported similar developments for monocrystalline silicon in which 156×156 mm cells were processed with all-screen-printed contacts and an efficiency of 20.3% was achieved on 170 μm thick wafers. The rear aluminium paste was modified to prevent void formation in the rear contact [46]. Trina continued development, and in 2016 posted a record PERC efficiency of 22.6% ($243.2 \, cm^2$) for a monocrystalline solar cell and 21.3% for a multicrystalline silicon one [47], with average cell efficiencies in production of 21%. Research continues to be directed toward cost reduction by improving efficiency and throughput. Production rates of 3600 wafers per hour are achievable [48]. Recombination in the emitter has been shown to be a major source of loss. Etching back of the emitter successfully reduced emitter saturation current densities from 90 to 22 fA/cm^2, indicating that 22% efficiency in production is achievable [49]. Selective emitters have also been successfully incorporated into the PERC structure [50].

Complete metal coverage of the rear surface is not essential, and therefore a rear grid contact allows light into the rear surface so that bifacial action of PERC cells in a module is possible [51]. This is the PERC+ design and is discussed in more detail later.

While present commercial exploitation is centred around PERC manufacture, intensive R&D effort continues in PERL and PERT devices, particularly for n type wafers.

8.2.4 Industrial Manufacture of PERC Cells

The IRTPV roadmap estimated that 30% of global silicon solar cell production would be of the PERC/PERT/PERL type and only 25% of the Al/BSF type by 2016 [52]. Actual expansion in PERC manufacturing capacity has happened rapidly. PERC capacity in China was estimated at 15 GWp in 2016 and had reached 92 GWp by 2019 [53]. All the major Chinese manufacturers are offering PERC-type cells

Building on Trina's reported 22.6% efficiency in 2016, other manufacturers have been posting claims for world-record efficiencies. Jinko Solar achieved 23.45% efficiency in December 2017 [54], while LONGi reported 24.06% efficiency in January 2019, with a commercial p type CZ wafer of 245.7 cm^2 [55]. Market studies project the growth of PERC technology up to 158 GWp p.a. by 2022 [56]. The driving forces behind this are efficiency-lowering installation costs and the deployment of photovoltaics in rapidly growing markets like Taiwan, where space is very limited. Table 8.2 lists the highest-efficiency commercial modules available in January 2020 for 60 cell modules. A clear hierarchy can be seen based on cell type and manufacturing technology. The highest-efficiency solar cells (and hence modules) are the n type interdigitated back-contact (IBC) cells and heterojunction solar cells (SHJs or HJCs); these are discussed in detail later. As Table 8.2 shows, new approaches to module assembly have been implemented in production. With increases in cell size and efficiency, currents have become large (over 11 A), so that resistive losses in the cell interconnection have become very significant. These can be reduced by using half cells in the module [57]. In addition, new interconnection methods have been developed. Shingling of solar

Table 8.2 Most efficient solar modules (60-cell) in 2020, by manufacturer and cell type [57]

Manufacturer	Power (Wp)	Cell type	Efficiency (%)
Sunpower	400	n type IBC	22.6
LG	375	n type IBC	21.7
REC Solar	380	HJC n type half cell	21.7
LG	355	n type mono	20.7
Sunpower	360	n type IBC	20.4
Longi Solar	355	p type mono PERC shingled	20.3
Hanwha Q Cells	350	p type mono PERC half cell	20.1
REC Solar	330	n type mono half cell	19.8
Jinko	330	p type mono PERC half cell	19.6
Trina Solar	330	p type mono PERC half cell	19.6
JA Solar	330	p type mono PERC half cell	19.6
Sumec Phono Solar	330	p type mono PERC half cell	19.6

Source: Modified from www.cleanenergyreviews.info/blog/2017/9/11/best-solar-panels-top-modules-review accessed 5/3/2020

cells, although known about since the 1960s, has recently received attention due to the development of good electrically conductive adhesives. Rather than use interconnecting strips, cells are overlaid, with the rear edge of one overlapping the front edge of the next, eliminating the gap between them. This improves module efficiency [58]. An alternative approach is the use of the smart-wire connection technology (SWCT) developed by the Meyer Burger Company [59]. Rather than using the copper strip interconnect soldered to bus bars, multiple wires are bonded to the silicon surface, reducing the need for multiple printed bus bars on the cell surface. The SWCT reduces surface shading and increases current carrying capacity, leading to improvements in module efficiency and reducing silver consumption by eliminating the printed bus bar. LG, Trina, REC, and Canadian Solar have all adopted this approach [57].

Such is the drive for higher efficiency that the fraction of crystalline silicon production of monocrystalline wafers is actually increasing over that of the lower-cost multicrystalline option, as modules using multicrystalline cells are typically in the range of 16–17% efficiency [57].

8.2.5 Bifacial Module Technology

The world solar market in 2020 presents a number of challenges. The scale of manufacture means that most of the economies of scale have been achieved and the market remains very competitive, with many manufacturers at the gigawatt scale of production. Achieving further total system cost reduction is desirable in order to ensure cost-competitiveness against fossil-fuel electricity generation in all markets. Balance-of-system costs have followed module costs downward, but these are largely made up of other costs than the inverter (such as structures and cabling), which are unlikely to fall further. Therefore, it is essential to reduce system costs by achieving a higher energy output. The move to PERC technology has been part of this goal, as already described, but the manufacturing process is already much more complex than for the Al-BSF case in order to achieve a 1% absolute increase in efficiency. As described later, other approaches are underway which may further increase cell efficiency above 26%, but these require much more complex processing and a higher-quality wafer, and hence greater cost. A much quicker and simpler avenue toward increasing photovoltaic module output is the use of both sides of the module to generate power – hence the term 'bifacial'. When the back of the cell was covered in aluminium (as in the case of Al-BSF), bifaciality was not an option. With PERC cells, however, the metal on the rear can be in grooves with an H pattern matching the cell front, allowing light in from both sides [60]. This can be achieved without any increase in cell manufacturing cost. The modification is shown in Figure 8.7.

The bifacial concept is not new. The use of bifacial cells in concentrators was first proposed by Luque in 1978 [61], and in 1981 cells with rear side efficiency under illumination of 86% of the front side were demonstrated [62]. This bifacial technology was subsequently transferred to the Isofoton Company in Malaga, Spain, which

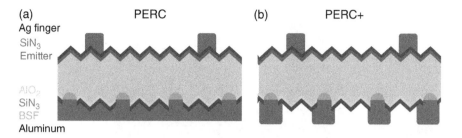

(a)   PERC   (b)   PERC+

Ag finger
SiN$_3$
Emitter

AlO$_2$
SiN$_3$
BSF
Aluminum

Figure 8.7 Conventional PERC cell with (a) full rear-surface aluminium coverage and (b) an aluminium grid on the rear, allowing bifacial light collection [60] (*Source:* T. Dullweber et al: Progress in Photovoltaics-Research and Applications 24, 1487-1498. © 2016, John Wiley & Sons.)

manufactured bifacial modules. The benefits were published [63], but here was little take-up of the technology for the next 20 years [64]. However, interest was greatly revived in 2015 as the PERC cell was established in large-scale manufacture and the potential for bifaciality was promoted [60]. Bifacial cells were demonstrated with a front-side efficiency of 20.8% and a rear-side one of 16.5%, with a bifaciality of 79%. The first bifacial PERC modules were introduced to the market by Solarworld at the same time [60]. It was claimed that they increased energy yield by 5–25%. The economic feasibility was investigated for a number of bifacial cell configurations in addition to PERC, and it was concluded in all cases that the bifacial cell gave a lower LCOE than an 18% multicrystalline Al-BSF cell in all cases with rear illumination less than 0.1 suns [65]. Solar cell development has continued with the addition of selective emitters to give a front-side efficiency up to 21.9% [66], and a module with efficiencies of 19.8% front side and 16.4% rear side has been independently verified [67]. Other cell types, such as SHJ [68] and n type PERT, have been developed as bifacial cells [69].

The critical thing in the adoption of the bifacial approach is the actual gain that can be achieved by collecting the albedo. This can vary enormously depending on the surroundings of the module and its reflectivity. With a highly reflecting white membrane over a concrete surface, the gain for a single module can be as high as 49% [70], but this is unreal. The best data are obtained from field testing data of large arrays. A range of 8–10% bifacial gain in energy has been measured for different locations in the United States [71], and a similar gain of 10% has been recorded in a large array in Chile [72]. Bifacial modules can be used in special applications, such as in noise barriers incorporating photovoltaics along motorways. The benefits are that electricity is generated in urban areas close to the point of use and the modules take up no additional space. The orientation is dictated by the course of the motorway, so bifacial modules have a definite advantage. An 8 kWp array at Aubrugg, Switzerland giving a yield of 681 kWhr/kWp is shown in Figure 8.8 [73].

These gains are credible enough that the deployment of bifacial modules is growing strongly. All the manufacturers in Table 8.2 are offering bifacial module products. The Trina Solar Duomax Twin bifacial module was launched in April 2017, and the

Figure 8.8 Bifacial 8.27 kWp photovoltaic noise barrier installed in Aubrugg, Switzerland [73] (*Source:* T. Nordmann and L. Clavadscher, 2004.PV on noise barriers. Progress in Photovoltaics, 12, 485-495. © John Wiley & Sons, Inc)

company had already installed 20 MWp of product [74]. The module is illustrated in Figure 8.9, showing the appearance from both sides.

8.2.6 Light-Induced Degradation

Chapter 7 highlighted the instability problems affecting many thin-film solar cell technologies. It was once generally assumed that crystalline silicon modules were stable under sunlight and that long-term ageing only occurred at around 0.5% of initial power per year. It was thus a considerable surprise when a rapid initial loss of output power for silicon modules was noticed in the 1990s, particularly for cells made with CZ monocrystalline silicon [75]. Although the light-induced degradation (LID) in boron-doped CZ silicon had been noted as long ago as 1973 [76], it was only in the 1990s as research effort focussed on high-efficiency approaches that the degradation became apparent [77], as higher efficiencies were critically dependent on maintaining a high minority carrier lifetime in the p type base. It was found that when the p type doping was as high as 0.5 Ω/cm, 7% degradation of initial efficiency took place. This degradation occurred rapidly – usually within a few hours of light exposure – but could be recovered by a low-temperature anneal at 200°C. Placing the cells in forward bias also induced it. The phenomenon was quickly identified as being due to the formation of an oxygen–boron complex [78], the effect of which is shown in Figure 8.10. It was observed that FZ silicon and magnetic CZ silicon with very low levels of oxygen showed no degradation and multicrystalline silicon with lower levels of oxygen than CZ silicon showed very little LID. To avoid the LID in CZ silicon, it became the norm

Figure 8.9   Trina Solar Duomax
Twin 60 bifacial glass–glass module
(*Source:* Trina Solar Ltd)

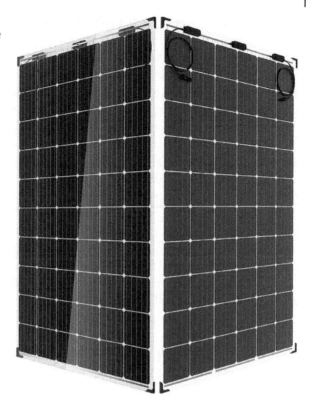

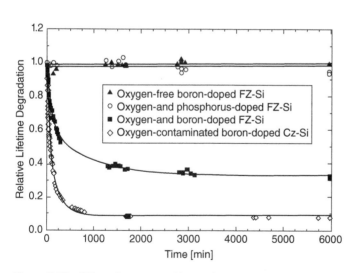

Figure 8.10   Effect of oxygen and boron doping on the relative loss of minority carrier lifetime
due to light exposure in p type silicon [77] (*Source:* T. Saitoh et al: Progress in Photovoltaics-
Research and Applications 8, 537-547. © 2000, John Wiley & Sons)

to use wafers with resistivities higher than $1\,\Omega/cm$, even though higher initial solar cell efficiencies were achieved with lower resistivities [75]. Other paths were also explored, with doping with gallium instead of boron shown to be particularly effective [79], although this was not widely adopted as a manufacturing technology due in part to the lower segregation coefficient of gallium in liquid silicon causing a much wider variation in resistivity along the CZ boule than in boron-doped growth, as well as the lack of a supply chain for gallium-doped feedstock.

In addition to the thermal annealing at $200\,°C$ used to stabilise the solar cells, it was found that a lower-temperature process with annealing at $70\,°C$ under illumination could permanently deactivate the defect. An anneal time over 100 hours was needed, although this was reduced at higher temperatures [80]. This was not acceptable as an industrial process, and instead manufacturers concentrated on fabricated cells with resistivities in the range $3–6\,\Omega/cm$. Recently, progress has been made in developing industrial equipment to stabilise the boron–oxygen complex in realistic times. A rapid-fire inline furnace with illumination heats the wafers to over $200\,°C$ for 40 seconds to stabilise the complex, with a total cycle time of 70 seconds. The process was demonstrated on commercial p type PERC cells and a relative change in efficiency of 1% was observed after 24 hours' light exposure, indicating removal of the LID [81].

An alternative approach has been to move away from boron-doped p type silicon and toward the development of n type silicon solar cells. In addition to the avoidance of LID, n type silicon offers a higher minority carrier lifetime than p type at the equivalent doping level.

8.3 Solar Cells with n Type Silicon

While the majority of commercial production remains p type, n type silicon offers advantages in terms of higher efficiency potential and the avoidance of LID. At the same resistivity, n type silicon has a higher minority carrier lifetime than p type. Although very early cells were n type, the p type wafers were more readily available and more radiation-hard for the satellite market and it was far easier to produce a shallow phosphorus n type emitter in p type silicon than to create a boron-doped p type emitter in n type. Much research has been directed to overcoming this difficulty.

8.3.1 SHJ Solar Cells

As noted in Chapter 7, the ideal combination of bandgaps for a two-cell tandem device is an upper cell with a bandgap of 1.8 eV and a lower cell with a bandgap of 1.1 eV, which corresponds very well to an amorphous silicon cell on a crystalline silicon lower cell. An early attempt to achieve such a structure as a heterojunction device rather than a tandem cell was made in 1983, when p type amorphous silicon was grown directly on multicrystalline silicon and a 9%-efficient solar cell was demonstrated [82]. This rather low efficiency was due to the high recombination at the interface between the p type amorphous silicon and the silicon substrate.

Significant progress was made when it was found that inserting an intrinsic amorphous silicon layer up to 10 nm in thickness gave a large improvement in solar cell efficiency, boosting both the Isc and the Voc. The solar cell type was designated as a heterojunction with intrinsic thin layer (HIT). HIT became a trademark for Sanyo, and later for Panasonic as it took over Sanyo. Initially, the intrinsic layer was applied only to the front surface, although n+ amorphous silicon was applied to the rear surface to form a BSF. An 18.1% (1 cm^2) efficiency was demonstrated with a textured monocrystalline silicon wafer [83]. The cell showed no degradation for light exposure up to 10 hours at 5 suns. Its structure is shown in Figure 8.11.

The high sheet resistance of the amorphous silicon layers made it necessary to apply a TCO for good current collection. A silver grid was printed on front and rear, although this was a low-temperature curing material, as high-temperature firing degraded the amorphous silicon passivation. It is apparent from Figure 8.11 that the HIT cell was inherently bifacial. Sanyo continued development by inserting an I layer amorphous silicon at the rear, as also shown in the figure, and further optimising the process. By 2000, a cell efficiency of 20.1% was achieved for a full-size 101 cm^2 cell, with a Voc over 700 mV, confirming the good passivation and the benefit of the heterojunction [84]. A further benefit of the high Voc was that the decrease in efficiency with temperature was much less than for a conventional silicon cell.

Sanyo started commercial manufacture of the HIT cell in October 1997, with an average module efficiency of 15.2%, which was very high for that time. A bifacial module was also included in the product range [84]. Improvement was made in reducing plasma damage to the silicon surface, and 21.5% efficiency was achieved in 2005 [85]. Further development addressed optimisation of the TCO and the metal contact, resulting in a record-efficiency HIT cell of 23.7% for a large-area 100.7 cm^2 cell on a CZ wafer only 98 μm thick [86]. A further optimisation gave an efficiency of 24.7% (102 cm^2), again on a thin wafer [87]. Research at Sanyo was then directed toward the potentially higher-efficiency rear-contact heterojunction solar cell, as described in the next section. A further advance in SHJ cell technology was made by Kaneka in using an

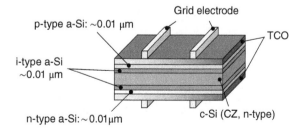

Figure 8.11 Structure of a HIT cell with n type silicon wafer and passivating intrinsic amorphous silicon layers (*Source:* Courtesy Panasonic Corp)

Grid electrode

p-type a-Si: ~0.01 μm

TCO

i-type a-Si ~0.01 μm

n-type a-Si: ~0.01 μm

c-Si (CZ, n-type)

electrodeposited copper contact at near room temperature, with an efficiency of 25.1% on a large-area 151.9 cm^2 cell independently verified [88].

Industrial production of the HIT cell continues, with Panasonic – in joint venture with Tesla – aiming for 1 GWp production by 2019 [89]. While Panasonic has been the dominant producer of HIT cells, other manufacturers such as GCL-Si have shown interest in them [90], while Meyer Burger Technology is offering turnkey production lines for the SHJ technology [91].

8.3.2 IBC-SHJ Solar Cells

As solar cell efficiencies approached the practical limit of 25%, interest focussed on the next step in development of the SHJ technology. The concept of the IBC cell was first proposed in 1977 [92] and is described in detail in the next section. In 2007, Birkmire identified the advantages of the IBC-SHJ approach [93]. The low-temperature processing of the SHJ technique meant very thin wafers could be processed without thermal stressing. Efficiency was increased by not having a top contact, and the contact area could be increased without shading the cell. Finally, the emitter did not have to transfer minority carriers, relaxing quality requirements. However, a high-minority-carrier-lifetime silicon wafer was required for good efficiency. Based on device modelling from these early experiments, it was predicted that the IBC-SHJ cell could achieve over 24% efficiency, although only a low efficiency was demonstrated in practice, albeit with a Voc of 691 mV [93]. A simple IBC-SHJ is shown in Figure 8.12.

Other research groups began working on this structure, and by 2010 a 15.7%-efficiency (25 cm^2) cell had been reported [94]. Rapid progress was made with the technology, and in 2014 Sharp reported a 25.1% (3.713 cm^2) cell efficiency [95], while at the same time Panasonic achieved an independently verified solar cell efficiency of 25.6% with a full cell area of 143.7 cm^2 – the world record for a single-junction silicon solar cell [96]. This record was overtaken by Kaneka in 2016 with a 26.3% (180.4 cm^2) cell [97], which followed the schematic structure of Figure 8.12, with a

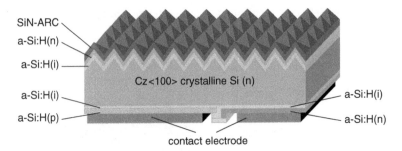

Figure 8.12 Structure of an IBC-SHJ cell [95] (*Source:* J. Nakamura et al: Proceedings - EU PVSEC 2014, 29th European Photovoltaic Solar Energy Conference and Exhibition, 373-375. © 2014, WIP GmbH & Co Planungs-KG. Courtesy EUPVSEC.)

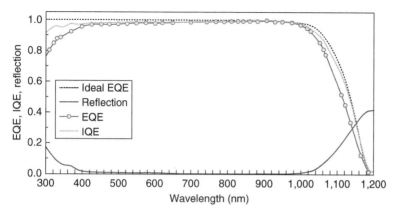

Figure 8.13 Quantum efficiency and reflectance spectra for the 26.3% HJ-IBC cell, as measured at FhG.ISE [97] (*Source:* Yoshikawa, K., Kawasaki, H., Yoshida, W., Irie, T., Konishi, K., Nakano, K., ... Yamamoto, K. (2017). Silicon heterojunction solar cell with interdigitated back contacts for a photoconversion efficiency over 26%. Nature Energy, 2(5), 17032. © 2017, Springer Nature. Courtesy Nature.)

$3\,\Omega/cm$ n type Cz wafer of $165\,\mu m$ thickness. The front surface was anisotropically textured and passivated with a-Si, and a double-layer antireflection coating was applied. The rear-surface heterojunctions were applied and patterned as i-p+ a-Si and i-n+ a-Si, with a design optimised to minimise series resistance. Very low surface reflectance was achieved and a Jsc of $42.3\,mA/cm^2$ was verified. The EQE, IQE, and reflectivity are shown in Figure 8.13. This result was overtaken by the same group in 2017 with a 26.7% ($79.0\,cm^2$) cell efficiency, which remains the world record for a silicon cell. At the same time, a record module efficiency of 24.4% was established [98]. The authors identified the potential for future improvement toward a 29% efficiency.

8.3.3 n Type IBC Cells Without Amorphous Silicon Passivation

The IBC-HJ cell is a relatively recent innovation, with the record efficiency for an IBC cell held for many years by the SunPower Corporation's high-efficiency devices. The early history has been described by Swanson [99]. Initial interest stemmed from high-efficiency cells designed for thermo-photovoltaic (TPV) generation in which sunlight was focussed on to an absorber which reradiated to a silicon solar cell. This was quickly displaced by a simple concentrator design, with a 20%-efficient solar cell demonstrated in 1982. By 1985, a 25% efficiency was demonstrated at 200 suns with a cell area of $0.15\,cm^2$. The best efficiencies were obtained with an n type float-zone wafer of 100–$200\,\Omega/cm$ and thicknesses between 100 and $160\,\mu m$ [100]. The front and rear surfaces were passivated with silicon oxide. N+ and p+ point diffusions typically of $100\,\mu m^2$ were made on the rear surface and connected in the correct polarity by bus bars. A schematic is given in Figure 8.14.

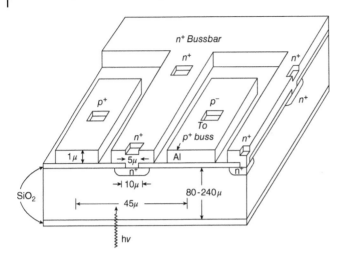

Figure 8.14 Cross-section of a rear-contact solar cell with point diffusions [100] (*Source:* R.A. Sinton et al: Proceedings of the 18th IEEE Photovoltaic Specialists Conference 61-65. © 1985, IEEE. Courtesy IEEE.)

In addition to the front oxide, a double-layer ARC was applied. Several photolithographic steps were necessary to produce the cell, and maintenance of a high-minority-carrier-length wafer was crucial to achieving high efficiency. The cell was expensive and could only be economically used in concentrator applications. Modelling showed that with a minority carrier lifetime of 3 ms, an efficiency of 28% at 1500 suns was achievable [101]. In 1986, a cell efficiency of 27.5% at 100 suns was reported [102]. The original work was done by Swanson and coworkers at Stanford University. In 1990, there was sufficient confidence in the IBC approach for commercialisation, and the SunPower company was established [99]. By 2000, it was was offering production concentrator cells of 26% efficiency under concentration [103], in addition to 1 sun cells of around 22% efficiency for solar race cars and the NASA Helios project.

However, it was apparent that commercial take-off of concentrator systems was not happening and development efforts were directed to achieving very high efficiency for flat-plate applications [99]. This move was prompted by the semiconductor manufacturer, Cypress, taking a majority stake in the company. By 2003, a nonphotolithographic process had been developed and the point diffusions were replaced by stripes with screen-printed contacts [104]. A best cell efficiency of 20.4% was measured by NREL for a 148.8 cm^2 n type float-zone wafer. The design still required a minority carrier lifetime in excess of 1 ms to allow minority carriers to travel across the wide diffusions, which were greater than the wafer thickness. The design was relatively insensitive to wafer thickness, and a resistivity range of 2–10 Ω/cm was acceptable. By 2006, SunPower was manufacturing cells at 75 MWp p.a. with an average efficiency in production of 20.6% and a best cell efficiency of 22.1% [105]. Development continued, with SunPower reporting a world-record efficiency of 24.2% (155.1 cm^2) [106] and a module efficiency of 21.4% [107] in 2010. The company built up manufacturing capacity, first at its plant in the Philippines and then in Malaysia. In 2011, majority ownership passed to the Total Oil group. Currently, the total manufacturing capacity

Figure 8.15 A typical SunPower installation with bifacial modules tracked to enhance output. Source: M P Campbell et al Proceedings 22ndEU PVSEC (2007) 22nd EuropeanPhotovoltaic Solar Energy Conference and Exhibition 976–979 © 2007WIP GmbH &Co Planungs-KG

of all types is 1.9 GWp, and the capacity for the highest-efficiency product is around 350 MWp p.a. [108]. Additional manufacturing facilities are planned at the former SolarWorld facility in Oregon, United States. SunPower currently offers its best modules with an average total area efficiency over 22% (Table 8.2) and has demonstrated a 25%-efficient cell in the laboratory [109]. Figure 8.15 shows high power SunPower modules in the field.

In addition to SunPower, other companies and institutions have developed IBC cells on n type wafers. Trina Solar, initially in collaboration with the Australian National University, demonstrated a 4 cm^2 cell with an efficiency of 24.4% using photolithographic processes [110]. It subsequently simplified the process to use screen printing and selective emitters, and demonstrated a 24.13% (244.9 cm^2) full-size cell, although this is only at the pilot stage [111]. The structure is shown schematically in Figure 8.16. An alternative structure was developed at the International Solar Energy Research Centre (ISC, Konstanz, Germany), the Zebra cell, in which tube diffusion is used for boron and phosphorus diffusions. A front floating emitter is used with silicon oxide/silicon nitride passivation for the front and rear. Metallisation is by laser ablation and fire through screen-printed contacts. The best cell result achieved is 21.9% (239 cm^2) efficiency in pilot production [112]. Further commercial exploitation is planned.

One advantage of the IBC approach is that it does not require the soldering of the contact tabbing from the front side of one cell to the rear of the next to create the series string. If all the contacts are on the rear, planar interconnection is possible; this is a

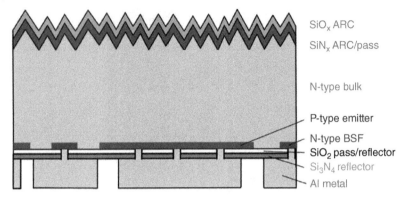

SiO$_x$ ARC
SiN$_x$ ARC/pass

N-type bulk

P-type emitter
N-type BSF
SiO$_2$ pass/reflector
Si$_3$N$_4$ reflector
Al metal

Figure 8.16 Two-dimensional schematic diagram of an ANU/Trina Solar IBC cell [110] (*Source:* E. Franklin et al: Progress in Photovoltaics-Research and Applications 24, 411-427. © 2016, John Wiley & Sons.)

particular benefit when thinner wafers are processed. Potentially, costs can be reduced further if a prepatterned conductive back sheet is used, with the cells placed on the back sheet and electrical contact made by an electrically conductive adhesive, which cures in the lamination process. Good results have been obtained using this approach, with cell-to-module losses of only 1% power [113]. The absence of bus bars on the cell front coupled with the black backsheet gives a very uniform appearance, making the module well suited for BIPV applications. Although the high efficiency reduces overall system costs, module price is high at approximately twice that of the lowest-cost product [114].

8.4   The Future of Photovoltaic Technology: Toward Terrawatts

The technologies described in the previous section have enabled a photovoltaics market at the scale of 0.1 TWp to be established. The impacts of climate change require significant changes in global energy supply to avoid devastating consequences. This has been modelled by the International Renewable Energy Agency (IRENA) in its annual report on the Global Energy Transformation [115], which proposes that by 2050, 50% of global energy supply will be from electricity, including electric cars for transport and heat pumps for domestic heating, and that 86% of this electricity supply will come from renewables, with photovoltaics making up the major share. To meet this target, 8.5 TWp of photovoltaics will have to be installed by 2050, up from the 0.5 TWp in place in 2018. This implies a large increase in annual shipments from the present 100 GWp p.a., and the IRENA report postulates 350 GWp annual shipments in 2050. Achieving this requires the levelised cost of energy (LCOE) for photovoltaics to be $38/MWhr, which doesn't seem unrealistic given the

PPAs already in place in 2019, as described in Chapter 6. This has clear consequences for the future of photovoltaics manufacturing and the cell technologies that will be employed. Hoffmann and Metz concluded that based on the PERC technology for monocrystalline silicon wafers and the application of learning-curve theory, module cost would reduce to $0.1/Wp for a 10 TWp total global installation by 2050 [116]. Although based on very realistic projections, their assumed solar cell efficiency was 25.4%, which seems low compared with current cell efficiencies being demonstrated at over 24% on full-size wafers [106]. While the limit for single-junction silicon solar cell efficiency is being reached, it is highly likely that progress beyond the 25% barrier in production will be made in the next 30 years, with 30% tandem cell efficiency identified as a viable target [117]. Given the high level of investment in silicon manufacturing technology and the good track record of the performance of silicon photovoltaics, it also seems probable that future manufacturing technology will evolve as an add on to the existing silicon processes.

The use of tandem cells to enhance solar cell efficiency is well known and has been extensively used for amorphous and microcrystalline silicon, as described in Chapter 7. Tandem cells are the cornerstone of development in III–V cells for space and concentrator applications, but the approach has not been used for crystalline silicon. The two most prospective options for tandems on silicon are for III–Vs and the emerging perovskite.

8.4.1 III–V Tandems on Silicon

The very high efficiency capability of III–V solar cells based on GaAs and its alloys is well known, but is limited for terrestrial applications by the high cost of substrate and cell processing. On the other hand, silicon offers a low-cost substrate with a multigigawatt-scale manufacturing capability. This is a promising route to very high efficiencies at low costs. The bandgaps of silicon at 1.1 eV and of the III–V alloys at 1.4–2 eV offer an ideal combination for high-efficiency tandem solar cells. The potential for 38% efficiency has been identified [118]. Unfortunately, the 4% lattice mismatch between silicon and GaAs means it is very difficult to grow good-quality III–V material on to silicon. This is a challenge that researchers have been attempting to address since the early 1980s [119], but direct epitaxial growth remains problematic [120]. The lattice mismatch causes dislocations to be generated at the silicon/III–V interface and propagate through the epitaxial layer. These dislocations are active recombination centres which significantly reduce solar cell efficiency. The best result reported for III–V solar cells grown directly on to silicon is an efficiency of 24.3% (4 cm^2) for a GaInP/GaAs/Si structure [121].

Other approaches that do not use epitaxial growth have been developed. Simply mechanically stacking a III–V cell above a silicon one, with both operating independently, achieved an efficiency of 35.9% (1 cm^2) [122]. However, the result was a four-terminal device and thus not practical. Alternatively, it is possible to bond cells together, as shown in Figure 8.17 [123]. Here, the III–V tandem is grown by

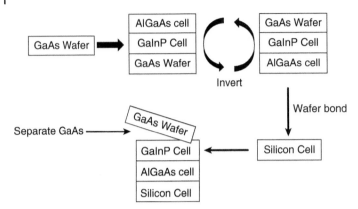

Figure 8.17 Schematic process for assembling a triple-junction solar cell on silicon [123] (*Source: Modified from Cariou, R., Benick, J., Beutel, P., Razek, N., Flotgen, C., Hermle, M., ... Dimroth, F. (2017). Monolithic Two-Terminal III–V//Si Triple-Junction Solar Cells With 30.2% Efficiency Under 1-Sun AM1.5g. IEEE Journal of Photovoltaics, 7(1), 367–373.*)

conventional MOCVD and the silicon bottom cell is made by phosphorus ion implantation into silicon. This process was carried out on a 4 inch silicon wafer, which was successfully bonded to the III–V tandem cell, with the GaAs substrate removed by etching a sacrificial layer between it and the active cell. Such removal is a well-established technique. Initially, a 30.2%-efficient solar cell was achieved on a 4 cm^2 device cut out from the larger wafer [123]. This result was improved further to achieve 33.3% solar cell efficiency [124] – the current best result for a tandem silicon/III–V cell [125].

While recognising the achievement in passing the 30% cell-efficiency milestone, the full costs of III–V cell processing remain high [122] and it has been argued that the tandem silicon/III–V cell will never be cost-effective except in the most demanding area-dependent applications [126]. The long-term challenge therefore is to develop low-cost processing procedures for III–V cells.

8.4.2 Silicon Tandems Using Perovskites

In contrast to the III–V alloys, perovskites offer the potential for a very low-cost solar cell technology based on room-temperature methods such as printing and spinning, as discussed in Chapter 7. High-efficiency silicon cells with an IBC structure have a suitable surface for deposition of the perovskite cell, and a number of research groups are actively investigating this potential. A first proof of concept was obtained in 2015 for a monolithic perovskite on an SHJ cell, as shown in Figure 8.18 [127]. An 18% solar cell efficiency was achieved.

Optical modelling has shown the good match between perovskite and silicon in achieving a tandem solar cell efficiency over 30%, assuming a perovskite bandgap of 1.7 eV and a thickness of 1.5 μm [128]. The absorption spectrum for an optimised tandem is shown in Figure 8.19.

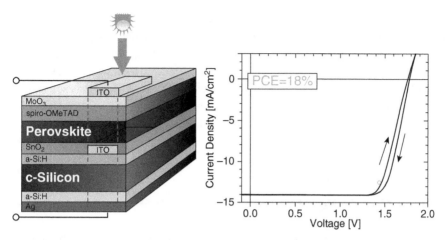

Figure 8.18 Proof-of-concept monolithic perovskite on an SHJ solar cell [127] *Source:* Albrecht, S., Saliba, M., Correa Baena, J. P., Lang, F., Kegelmann, L., Mews, M.,... Rech, B. (2016). Monolithic perovskite/silicon-heterojunction tandem solar cells processed at low temperature. Energy & Environmental Science, 9(1), 81–88. © 2016, Royal Society of Chemistry.

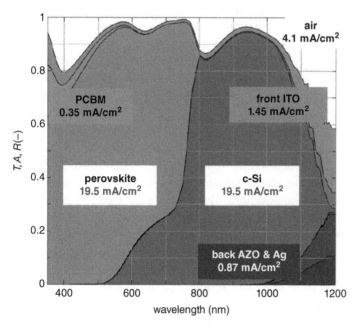

Figure 8.19 Absorption profile of an optimised perovskite silicon tandem cell with texture-etched rear silicon surface [128] (*Source:* K. Jaeger et al: Proceedings - EU PVSEC 2017, 33th European Photovoltaic Solar Energy Conference and Exhibition, 1057-1060. © 2017, WIP GmbH & Co Planungs-KG. Courtesy EUPVSEC.)

Good progress has been made since the first proof of concept. Two architectures have been followed: a monolithic structure, as shown in Figure 8.18, in which there are only two terminals, following conventional top and bottom contacts; and a four-terminal device, in which the output from the two cells is collected separately but the cells are connected optically. In 2017, a four-terminal perovskite/silicon device gave 24.1% efficiency [129]. More recently, the 25% efficiency barrier was passed by two groups, with CSEM/EPFL reporting a 25.2% efficiency [130] and IMEC passing the single-junction silicon solar cell record with 27.1% efficiency on a small area ($0.13\,cm^2$), reducing to 25.2% on $4\,cm^2$ for a four-terminal device [131]. This record was surpassed in turn by Oxford PV, with a 28.0% efficiency on $1\,cm^2$ verified by NREL [132]. Oxford PV is developing this technology on its $17\,000\,m^2$ pilot line in Germany.

The approach is still at an early stage, and the commercialisation prospects are uncertain.

8.5 Silicon Module Reliability

Part of the reason for the dominance of the silicon-based photovoltaics technologies has been a perception of their high reliability in use and their freedom from the degradation modes affecting the thin-film technologies and the emerging polymer-based technologies, including perovskites. They have now been used in significant numbers for over 35 years. Evaluation of their long-term reliability has been undertaken. In 2007, the European Solar Test Installation reported its findings on the 20-year life of a range of early modules, concluding that 82% showed less than 20% loss of maximum power after outdoor exposure in Northern Italy, consistent with manufacturers' warranties at the time [133]. More recent studies have distinguished between module failure modes and long-term degradation mechanisms. In general, failure rates are low at <0.1% [134].

The principal causes of module failure are shown in Figure 8.20. It can be seen that encapsulant discolouration and delamination have traditionally been major modes of failure (Figure 8.20a), but these have decreased significantly in the past 10 years as encapsulation materials and techniques have improved (Figure 8.20b). Of concern is the growth in potential induced degradation (PID). This is seen as a power loss in modules installed in large utility-scale applications where there is a large voltage between the encapsulated solar cells and their frame. It was noted in early SunPower modules, but reversing the polarity removed the problem [99]. However, the problem is of a more general nature, being a function of both the conductivity of the EVA encapsulant and the solar cell design [135]. Simulated PID stress tests have shown that losses of power up to 40% can occur, although these may be reversible depending on the module polarity [136]. The conditions under which PID occurs are becoming better understood, and module manufacturers are now prepared to guarantee PID-free operation [137].

8.6 Summary

This chapter has described the development of cell design and manufacturing techniques for crystalline silicon solar cells and modules. Crystalline silicon modules account for around 90% of global production, and their market share is increasing.

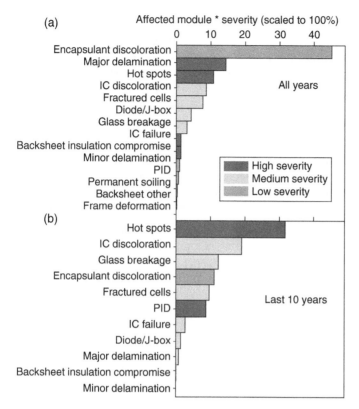

Figure 8.20 Relative severity of different modes of module failure as a percentage of total failures [134] (*Source:* D.C. Jordan et al: Progress in Photovoltaics-Research and Applications 25, 318-326. © 2017, John Wiley & Sons.)

Some of the reasons for this are highlighted in this chapter. While thin-film technologies have improved and champion cells are around 22%, there is still an absolute 5% advantage of the best crystalline silicon devices. Crystalline silicon has proven its ability to be manufactured in very high volumes up to nearly 100 GWp p.a., with year-on-year cost reduction. The efficiency of both R&D and production modules continues to improve. PERC is becoming established as the main silicon manufacturing technology and is applicable to both monocrystalline and multicrystalline wafers, with a world-record large-area multicrystalline PERC cell efficiency of 22.8% being posted [125]. The surprising thing is how long it has taken for PERC to move from its first demonstration in 1989 to volume production today. This reflects the conservatism within the industry and the necessity for a technology which is going to be deployed for at least 25 years but which must be paid for up front to be proven to be reliable in that time frame. It is only when an existing manufacturing technology has reached its limit that newer technology is introduced. Ideally, the new technology should be additional to the existing one, given the billions of dollars' investment it represents.

The exciting thing for crystalline silicon is the amount of innovation in solar cell design that has yet to be transferred into industrial production. PERL and PERT cell types are as old as PERC but have not yet been commercialised. SHJ, IBC, and IBC-SHJ cells are waiting to take their place as mainstream processes. The use of passivated contacts such as the TOPCon concept is emerging [138]. Even the route to 30% efficiency and beyond is being signposted by the emergence of the silicon tandem, whether with III–V alloys or – rather more probably – with a perovskite wide-bandgap cell. But the achievement of ever higher efficiencies is not the goal; the goal is a lower LCOE. The potential for reduced manufacturing costs is limited, as the cost of polysilicon to the photovoltaics industry has stabilised at around $10/kg for high volumes and all the economies of scale in manufacture have been achieved (although incremental improvement will surely occur). Increasing solar cell efficiency will play a role, but so too will new concepts such as improved cell interconnection and module encapsulation techniques, and the adoption and optimisation of bifacial modules offers a secure route to a higher-yield electricity generation. Module lifetime is also relevant here. Most LCOE costing is based on a 20-year module productive life. However, a 25-year working life is now guaranteed, 35 years is seen as achievable in the medium term, and researchers are looking ahead to a 50-year lifetime somewhere down the road. Amortising investment cost over these periods has a major impact on LCOE calculation.

The ongoing transformation to a carbon-free energy world has created an expectation that photovoltaics deployment at the terawatt level is needed. The crystalline silicon manufacturing technology is well placed to meet this challenge.

References

1 IHS Market report 5 April 2018.
2 International Technology Roadmap for PV 9th edn (2018).
3 M.A. Green in 'Solar Cell Materials – Developing Technologies' eds G.J. Conibeer and A. Willoughby pub: Wiley (2014) 65–84.
4 Photovoltaics Report FhG.ISE 19 June 2018.
5 M.A. Green et al.: Progress in Photovoltaics – Research and Applications 26 (2018) 3–12.
6 R.M. Swanson: Proceedings of the 31st IEEE PVSC (2005) 889–894.
7 S.W. Glunz: Proceedings of the 27th EUPVSEC (2012) 2BP.12.
8 S.R. Wenham: 'Laser Grooved Silicon Solar Cells' PhD thesis, University of New South Wales, Sydney, NSW, Australia (1986).
9 N.B. Mason et al.: Proceedings of the 10th EUPVSEC (1991) 280–284.
10 S. Eager et al.: Proceedings of the 29th IEEE PVSC (2002) 62–65.
11 A. Alonso-Abella et al.: Proceedings of the 29th EUPVSEC (2014) 2728–2733.
12 T.M. Bruton et al.: Proceedings of the 12th EUPVSEC (1994) 531–532.

13 N.B. Mason et al.: Proceedings of the 21st EUPVSEC (2006) 521–523.

14 T.M. Bruton et al.: Proceedings of the 29th IEEE PVSC (2002) 1366–1368.

15 T.M. Bruton et al.: Proceedings of the 23rd IEEE PVSC (1993) 1250–1251.

16 K.C. Heasman et al.: Proceedings of the 21st EUPVSEC (2006) 2197–2200.

17 H. Mughal and T.M. Bruton: Proceedings of the 24th EUPVSEC (2009) 774–776.

18 M. McCann et al.: Progress in Photovoltaics – Science and Applications 16 (2008) 467–477.

19 A. Lennon et al.: Progress in Photovoltaics – Science and Applications 27 (2019) 67–97.

20 A. Blakers and N. Zin in 'Photovoltaic Solar Energy – From Fundamentals to Applications' eds A. Reinders, P. Verlinden, W. Van Sark, and A. Freundlich pub: Wiley (2017) 84–85.

21 Feng Ye et al.: Proceedings of the 29th EUPVSEC (2014) 1069–1071.

22 T.C. Roeder et al.: Progress in Photovoltaics – Research and Applications 18 (2010) 505–510.

23 H.H. Kuehnlein et al.: Proceedings of the 24th EUPVSEC (2009) 1712–1714.

24 A. Voltan et al.: Progress in Photovoltaics – Research and Applications 20 (2012) 670–680.

25 W. Wu et al.: Progress in Photovoltaics – Research and Applications (2018) doi: 10.1002/pip.30.

26 M.B. Spitzer et al.: Proceedings of the 17th IEEE PVSC (1984) 398–402.

27 M.A. Green et al.: Proceedings of the 18th IEEE PVSC (1985) 39–42.

28 T.M. Bruton et al.: Proceedings of the 17th EUPVSEC (2001) 1282–1286.

29 A.W. Blakers et al.: Applied Physics Letters 55 (1989) 1363.

30 M.A. Green et al.: Proceedings of the 21st IEEE PVSC (1990) 207–210.

31 A. Aberle et al.: Progress in Photovoltaics – Research and Applications 1 (1993) 133–143.

32 J. Zhao et al.: Progress in Photovoltaics – Research and Applications 7(1999) 471–474.

33 M.A. Green et al.: Progress in Photovoltaics – Research and Applications 22 (2014) 1–9.

34 S.W. Glunz et al.: Proceedings of the 14th EUPVSEC (1997) 392–395.

35 E. Schneiderloechner et al.: Progress in Photovoltaics – Research and Applications 10 (2002) 29–34.

36 G. Agostinelli et al.: Proceedings of the 20th EUPVSEC (2005) 647–650.

37 M. Kerr et al.: Progress in Photovoltaics – Research and Applications 8 (2000) 529–536.

38 P. Choulat et al.: Proceedings of the 22nd EUPVSEC (2007) 1011–1014.

39 G. Agostinelli et al.: Proceedings of the 20th EUPVSEC (2005) 942–945.

40 M. Hofmann et al.: Proceedings of the 23rd EUPVSEC (2008) 1704–1707.

41 J. Schmidt et al.: Proceedings of the 23rd EUPVSEC (2008) 974–981.

42 P. Engelhart et al.: Proceedings of the 26th EUPVSEC (2011) 821–826.

43 P. Engelhart et al.: Energy Procedia 8 (2011) 313–317.

44 R. Sastrawan et al.: Proceedings of the 28th EUPVSEC (2013) 1861–1866.

45 B. Tjahjono et al.: Proceedings of the 28th EUPVSEC (2013) 775–779.

46 D. Chen et al.: Proceedings of the 29th EUPVSEC (2014) 1383–1386.

47 Trina Solar press release 18 December 2016.

48 F. Souren et al.: Proceedings of the 33rd EUPVSEC (2017) 967–969.

49 T. Dullweber et al.: Progress in Photovoltaics – Research and Applications 25 (2017) 509–514.

50 W. Wu et al.: Progress in Photovoltaics – Research and Applications (2018) doi: 10.1002/pip.3013.

51 T. Dullweber et al.: Proceedings of the 33rd EUPVSEC (2017) 649–656.

52 'International Technology Roadmap for PV' 7th edn (2016).

53 C. Lin: PV Magazine 24 September 2019.

54 Photon International December 2017.

55 Longi press release 16 January 2019.

56 M. Willhun: PV Magazine International 7 September 2018.

57 https://www.cleanenergyreviews.info/blog/2017/9/11/best-solar-panels-top-modules-review accessed 7 August 2020.

58 N. Klasen et al.: Proceedings of the 7th Workshop on Metallization and Interconnection of Silicon Solar Cells (2017) doi: 10.2139/ssrn.3152478.

59 https://www.meyerburger.com/user_upload/dashboard_news_bundle/376409e022f7d2ae6f6e29318f8055410774c7fd.pdf accessed 7 August 2020.

60 T. Dullweber et al.: Progress in Photovoltaics – Research and Applications 24 (2016) 1487–1498.

61 A. Luque et al.: Solid State Electronics 21 (1978) 793.

62 J. Eguren et al.: Proceedings of the 15th IEEE PVSC (1981) 1343–1348.

63 J. Perez et al.: Solar and Wind Energy 5 (1988) 629–636.

64 W. Koch et al.: 'Handbook of Photovoltaic Science and Engineering' eds A. Luque and S. Hegedus pub: Wiley (2003) 294.

65 F. Fertig et al.: Progress in Photovoltaics – Research and Applications 24 (2016) 800–817.

66 W. Wu et al.: Progress in Photovoltaics – Research and Applications 26 (2018) 752–760.

67 T. Dullweber et al.: Proceedings of the 33rd EUPVSEC (2017) 649–656.

68 W. Favre et al.: Proceedings of the 33rd EUPVSEC (2017) 437–440.

69 D. Lia et al.: Proceedings of the 33rd EUPVSEC (2017) 424–427.

70 M. Chodiodetti et al.: Proceedings of the 32nd EUPVSEC (2016) 1449–1552.

71 J.S. Stein et al.: Proceedings of the 33rd EUPVSEC (2017) 1961–1967.

72 E. Cabrera et al.: Proceedings of the 32nd EUPVSEC (2016) 1573–1578.

73 T. Nordmann and L. Clavadscher: Progress in Photovoltaics – Research and Applications 12 (2004) 485–495.

74 Trina Solar press release 1 April 2017.

75 S. Sterk et al.: Proceedings of the 14th EUPVSEC (1997) 85–87.

76 H. Fischer and W. Pschunder: Proceedings of the 10th IEEE PVSC (1973) 404–411.

77 T. Saitoh et al.: Progress in Photovoltaics – Research and Applications 8 (2000) 537–547.

78 J. Schmidt et al.: Proceedings of the 27th IEEE PVSC (1997) 13–18.

79 S.W. Glunz et al.: Progress in Photovoltaics – Research and Applications 8 (2000) 237–240.

80 A Herguth et al.: Proceedings of the 31st EUPVSEC (2015) 530–537.

81 D.C. Walter et al.: Proceedings of the 33rd EUPVSEC (2017) 377–381.

82 Y. Hamakawa et al.: Applied Physics Letters 43 (1983) 644.

83 K. Wakisaka et al.: Proceedings of the 22nd IEEE PVSC (1991) 887–892.

84 M. Taguchi et al.: Progress in Photovoltaics – Research and Applications 8 (2000) 503–513.

85 M. Taguchi et al.: Progress in Photovoltaics – Research and Applications 13 (2005) 481–488.

86 T. Kinoshita et al.: Proceedings of the 26th EUPVSEC (2011) 871–874.

87 R. Brendel et al.: Proceedings of the 28th EUPVSEC (2013) 676–690.

88 D. Adachi et al.: Applied Physics Letters 107 (2015) 233506.

89 https://news.panasonic.com/global/press/data/2016/12/en161227-6/en161227-6.html accessed 7 August 2020.

90 L. Zhao et al.: Frontiers in Energy 11 (2017) 85–91.

91 https://www.meyerburger.com/en/hjt-and-smartwire/ accessed 7 August 2020.

92 M. Lammert and R. Schwartz: IEEE Transactions on Electron Devices 24 (1977) 337.

93 M. Lu, S. Bowden, and R. Birkmire: Proceedings of the 22nd EUPVSEC (2007) 924–927.

94 T. Desrues et al.: Proceedings of the 25th EUPVSEC (2010) 2374–2377.

95 J. Nakamura et al.: Proceedings of the 29th EUPVSEC (2014) 373–375.

96 K. Masulo et al.: IEEE Journal of Photovoltaics 4 (2014) 1430–1435.

97 K. Yoshikawa et al.: Nature Energy 2 (2017) 17032.

98 M.A. Green et al.: Progress in Photovoltaics – Research and Applications 25 (2017) 3–13.

99 R.M. Swanson in 'Power for the World' ed. W. Palz pub: Pan Stanford (2014) 227–256.

100 R.A. Sinton et al.: Proceedings of the 18th IEEE PVSC (1985) 61–65.

101 R.A. Sinton and R.M. Swanson: Proceedings of the 19th IEEE PVSC (1987) 1201–1208.

102 R.A. Sinton et al.: IEEE Transactions on Electronic Device Letters EDL-7(10) (1986) 567–569.

103 R.M. Swanson: Progress in Photovoltaics – Research and Applications 8 (2000) 93–111.

104 K.R. McIntosh et al.: Proceedings of the 3rd World Conference on Photovoltaic Energy Conversion (2003) 4O-D10-05.

105 W.P. Mulligan et al: Proceedings of the 21st EUPVSEC (2006) 1301–1302.

106 P.J. Cousins et al.: Proceedings of the 35th IEEE PVSC (2010) 275–278.

107 M.A. Green et al.: Progress in Photovoltaics – Research and Applications 19 (2011) 84–92.

108 SunPower Annual Report 2017 NASDAQ-SPWR-2017.

109 D.D. Smith et al.: Proceedings of the 43rd IEEE PVSC (2016) 3351–3355.

110 E. Franklin et al.: Progress in Photovoltaics – Research and Applications 24 (2016) 411–427.

111 G. Xu et al.: Proceedings of the 33rd EUPVSEC (2017) 428–430.

112 G. Galbiati et al.: Proceedings of the 32nd EUPVSEC (2016) 980–983.

113 A. Halm et al.: Proceedings of the 32nd EUPVSEC (2016) 53–55.

114 https://www.pv-magazine.com/issue/01-2018/ accessed 7 August 2020.

115 IRENA: 'Global Energy Transformation – A Roadmap to 2050' (2019).

116 W. Hoffmann and A. Metz: Proceedings of the 36th EUPVSEC (2019) 1966–1971.

117 L. Oberbeck et al.: Proceedings of the 36th EUPVSEC (2019) 1950–1954.

118 S. Essig et al.: Energy Procedia 77 (2015) 464–469.

119 S.M. Vernon et al.: Proceedings of the 17th IEEE PVSC (1984) 434–439.

120 L. Vauche et al.: Proceedings of the 33rd EUPVSEC (2017) 1228–1231.

121 M. Feifel et al.: Solar RRL 3 (2019) 1903.

122 S. Essig et al.: Nature Energy 2 (2017) 171444.

123 R. Cariou et al.: IEEE Journal of of Photovoltaics 7 (2017) 367–373.

124 R. Cariou et al.: Nature Energy 3 (2018) 326–333.

125 M.A. Green et al.: Progress in Photovoltaics – Research and Applications 28 (2020) 3–15.

126 D.C. Bobela et al.: Progress in Photovoltaics – Research and Applications 25 (2017) 41–48.

127 S. Albrecht et al : Energy Environmental Science 9 (2016) 81–88.

128 K. Jaeger et al.: Proceedings of the 33rd EUPVSEC (2017) 1057–1060.

129 S.L. Luxemburg et al.: Proceedings of the 33rd EUPVSEC (2017) 1176–1180.

130 PV International News 41 (2018) 39.

131 PV International News 41 (2018) 78.

132 Oxford PV press release 28 December 2018.

133 A. Skoczek et al.: Proceedings of the 22nd EUPVSEC (2007) 2458–2466.

134 D.C. Jordan et al.: Progress in Photovoltaics – Research and Applications 25 (2017) 318–326.

135 A. Virtuani et al.: Progress in Photovoltaics – Research and Applications 27 (2019) 13–21.

136 W. Luo et al.: Progress in Photovoltaics – Research and Applications 26 (2018) 859–867.

137 https://www.q-cells.co.uk/products/technology/pid-lid.html accessed 7 August 2020.

138 Y. Chen et al.: Progress in Photovoltaics - Research and Applications 41 (2019) 46.

9

Lessons Learnt

9.1 Introduction

This book has charted how the photovoltaic technology has developed from the early days of the terrestrial market in the first commercial phase of the 1980s, where 30% of the global market was in low-power solar cells of a few milliwatts and less than 5% efficiency, to today's annual market of 100 GWp, with the largest installations over 1 GWp and the best solar cells having a cell efficiency above 20%. This has been accomplished in just over 30 years. It is a remarkable story in which many actors from inspired individuals to research teams large and small, major multinationals, governments, and supranational entities have played significant roles.

While Becquerel observed photoelectric effects in the 1820s and the first attempts to make working photovoltaic modules occurred in the 1880s, progress was very slow until the coming of the modern semiconductor age after 1948. The invention of the silicon solar cell was a direct consequence of the establishment of semiconductor expertise in the United States in the late 1930s and 1940s. As already detailed, the initial demonstration of a 5%-efficient silicon solar cell in 1954 led rapidly to the optimisation of the process and increased understanding, enabling the production of 10%-efficient cells. The silicon solar cell at that time was expensive and had no real application, and it was only the emergence of space satellites that gave photovoltaics a significant market and created confidence in the technology. Even then, photovoltaics might have remained a niche product with high added value, powering crucial military and telecommunication satellites. However, as detailed in Chapter 2, the energy crisis in 1973/74 stimulated the search for new, secure energy sources, and photovoltaics were identified as a viable technology in this regard, albeit one with major cost challenges. In hindsight, it is easy to see how priority should have been given to utilising an energy source that could deliver 6000 times the human population's energy demand, was available in viable quantities over most of the globe, and would last another 4.5 billion years. The reality was a long struggle over 30 years to demonstrate an economically viable electricity-generating technology, to create a

Photovoltaics from Milliwatts to Gigawatts: Understanding Market and Technology Drivers toward Terawatts,
First Edition. Tim Bruton.
© 2021 John Wiley & Sons Ltd. Published 2021 by John Wiley & Sons Ltd.

manufacturing base capable of supplying the global demand, and to deploy giga-watts of installations.

The journey is not over. Despite all the progress made, it is estimated that photovoltaics only provide 3% of our global electricity production [1]. While climate change was not an issue in the 1970s, it is a major concern now, and there is a recognition that in order to stabilise it at a 1.5 °C increase, net greenhouse gas emissions will have to fall to zero by 2050 [2]. There will thus be both an increasing demand for electricity and a need to reduce greenhouse gas emissions. Therefore, the demand for photovoltaics will continue to increase for the foreseeable future. This chapter looks at the roles that various actors have played in the past and how they must respond to the challenges of the future, as well as at how technology can be expected to evolve to meet these challenges.

9.2 Role of Governments

One of the main lessons learnt in the realisation of photovoltaics as a global energy source has been the vital role played by governments. The initial impetus, as described in Chapter 2, was the oil crisis in 1973/74. To that point, oil had been a major fuel for power generation. Suddenly, there was both a threefold increase in the oil price and an embargo by OPEC members on the sale of oil to the United States and some other countries. The immediate response of the US government was to insulate the country against both interruptions of supply and dramatic price increases. A wide range of alternative energy sources were thus investigated, one of which was photovoltaics. The challenge was to reduce the cost of the well-proven space technology, which had demonstrated that solar cells could work reliably for long periods without maintenance in the very arduous conditions of earth orbit, with high cosmic ray bombardment and extreme thermal cycling. Where the United States led, several other countries quickly followed – particularly Japan, which had a heavy dependence on imported energy, and Germany. Their governments' responses were to put in place long-term research programmes to reduce cost. Significant progress had been made by 1980, with the price of photovoltaics falling from the $100/Wp for space cells to below $10/Wp for terrestrial ones. At that price level, many off-grid applications became commercially viable, and by the early 1980s an embryonic manufacturing industry was in place, largely funded by wealthy multinationals, particularly in the oil industry.

Market stimulation was not given much focus at this point in time, beyond tax reliefs in some US states and capital grants in Japan and elsewhere, which as discussed were not effective in developing a market at the scale required to provide low costs. Growth occurred in the off-grid market, but infrastructure issues in less developed countries limited it. There was some funding for demonstration programmes, but again the volume was limited. Government R&D programmes remained strong, with an emphasis given to the development of thin-film photovoltaic technologies and, to a lesser extent, concentration solar systems. The 1980s saw new oil supplies coming on-stream, particularly from Alaska in the United States and the North Sea in Europe, and the oil price fell significantly. Development of renewables became of less concern to governments.

By the early 1990s, a new scenario was developing. While the cost of photovoltaics remained too high for central utility-scale generation, domestic- and commercial-scale grid connection was potentially viable in the near term. The Chernobyl nuclear disaster in the Soviet Union mobilised a strong anti-nuclear lobby, particularly in Germany [3], while the Iraqi invasion of Kuwait in 1990 and the subsequent First Gulf War reminded many of the political tensions around Middle Eastern oil supplies. The government response in Germany was the 1000 Roof Programme, as described in Chapter 6, which led to the introduction of feed-in-tariffs (FITs) and the explosive expansion of the solar market to achieve the global large-scale manufacture and cost-effective photovoltaic electricity generation seen in many countries today.

Ongoing roles for governments can thus be identified in the history of photovoltaic development. They played a key part in funding early R&D when it was by no means certain that a successful industry would develop. As described in Chapter 8, solar cell efficiencies of 26% have been achieved in the laboratory, and modules with solar cells of 24% efficiency are commercially available. While the ultimate efficiency for multijunction solar cells is theoretically close to 90%, there is considerable scope for future development. This will require investigation in basic science and long-term research, which is not readily funded by commercial operations. Thus, governments will need to continue to provide funding to universities and other advanced research institutes.

A further role for governments is to continue to ensure the photovoltaics industry has access to the electricity market. As this book has related, high-volume manufacturing has secured low costs for reliable products with >25-year service lifetimes, providing a low LCOE for new electricity generation – fully competitive with other forms of electricity generation, including fossil fuels. Nevertheless, photovoltaics remains a variable resource, and it is the function of governments to ensure that it has full access to national-grid networks and that small-scale producers at the domestic and commercial level can be properly remunerated for their contributions. It is of great importance if targets for greenhouse gas abatement are to be met that governments have policies in place to ensure that photovoltaic grid connection and self-consumption take place.

Governments have a further role in ensuring that only products which meet the various international standards regarding the quality, longevity, and safety of photovoltaic systems are implemented. This is important given that the main costs are in the initial capital investment for the system. Once installed, it should operate to its proposed energy yield, and should continue to do so for its designated lifetime, with the minimum of maintenance and unscheduled downtime. Each national government has the responsibility to see that there is a regulatory framework in place to guarantee these operational parameters for all photovoltaic systems.

9.3 Role of the Research Community

Clearly, the establishment of the global photovoltaic industry has been the outcome of a long research activity, from Becquerel's observation of the first instance of light creating a voltage, through the discovery of photoconductivity in solid selenium and the creation of the first working silicon solar cell at 5% efficiency, to today's world-record solar cell of 46% efficiency at 500× concentration. Photo-electricity was first observed in the nineteenth century, and by the early twentieth the role of electrons in conducting electricity and interacting with light was beginning to be understood. The challenge of making a solar cell with a meaningful output was not met until 1954, as described in Chapter 1. This discovery relied very much on the increased understanding of the materials science of semiconductors at the time, as well as the increased knowledge of how solar cells actually worked. It was rapidly refined in order to produce a silicon solar cell with radiation hardness capable of functioning well as a satellite power supply. The next challenge was to adapt the high-cost space cell technology for low-cost, high-volume manufacture. This was successfully achieved with the adoption of screen printing as the preferred technology for metal contact formation.

Through the 1980s, most of the concepts now used in the manufacture of high-efficiency silicon solar cells were developed, but these required maturity in the market place in order to be fully implemented, as described in Chapter 8. While silicon solar cells have been and remain the dominant manufacturing technology, alternative materials have been actively researched since the 1950s, although silicon's dominance has not been successfully challenged (see Chapter 7). Copper sulphide cells, amorphous silicon, and amorphous silicon/microcrystalline tandem cells were sequentially brought to the market, and each failed to achieve commercial success. The compound semiconductors CdTe and CIGS achieved gigawatt-scale production but remain a small segment of the market in comparison to silicon. Their large-area solar cell efficiencies lag behind those of silicon, and their potential advantages in a minimal-material-usage large-area monolithically integrated module have not been realised. Silicon continues to demonstrate increased solar cell efficiency and lower cost, so the future for thin films may lie in niche applications such as indoor power sources in the Internet of Things. Concentrating systems have demonstrated high system efficiencies, but difficulties in achieving low costs have not yet been overcome. The research task is not finished. The record silicon solar cell efficiency is 26%, which is close to the practical limit for a single-junction solar cell. The challenge is now to overcome this Schockley–Queisser limit [4].

At the end of the 1990s, it was postulated that there were three generations of solar cell technology. Silicon was the first generation, rather inaccurately typified as high-efficiency and high-cost. The second generation, which was expected to displace the first, was thin-film technologies, which were promoted as low-efficiency but low overall cost. Third-generation technologies were new approaches which would overcome the Schockly–Queissser limit through novel device physics, such as utilising intermediate-bandgap solar cells, hot electron capture, or multiple hole pairs from a single photon. Polymer solar cells and tandem devices were included in the third generation, although neither really fitted; tandem cells actually date back to the 1950s. So far, third-generation devices have not emerged as cost-effective solar cells. The short-term task is to find a cost-effective tandem solar cell structure to work with silicon as the bottom cell and take commercial solar cell efficiencies to beyond 30%. The two contenders, as outlined in Chapter 8, are III–V alloys and perovskites. Both present major obstacles to be overcome. Storage of electricity generated by photovoltaics is also a topic for future research, as discussed later.

9.4 Role of the Manufacturing Industry in Europe and the United States

It has been the role of manufacturers to create the commercial product, develop the markets, and install the systems. If the photovoltaic journey has been a success, then manufacturing and sales have been the most perilous part. They have been characterised by a high level of churn, with companies opening, flourishing, and disappearing. Although the Bell Company patented the first successful silicon solar cell and its offshoot Western Electric manufactured some of the early space cells, as described in Chapter 1, it soon ceased its activities. The Hoffman Semiconductor Company took up the charge but was soon overtaken by other companies, such as Spectrolab. The Sharp Corporation in Japan was an early entrant into the space cell market and later developed a terrestrial business, and is still in operation today. Before the oil crisis, Exxon funded its start-up solar cell company Solar Power Corporation in 1973. While it enjoyed considerable early success, it was closed in 1982. Solar Technology International was founded in 1975 and quickly became part of the Atlantic Richfield Company, trading as ARCO Solar. Through the 1980s, it was the world's largest photovoltaics manufacturing company. It was purchased by Siemens, sold to Shell, and finally became part of the Solar World Group. Production ceased at its historic Camarillo site in 2012 due to the introduction of low-cost modules from China. In 1973, Solarex was set up by staff leaving the COMSAT Laboratories. These were trailblazers, as the terrestrial photovoltaics market was small and fragmented in the 1970s. Solarex became part of the former American oil company Amoco in 1983.

These companies played a vital role in demonstrating that the manufacturing of terrestrial solar cells could be a meaningful business. International partnerships developed. Sharp developed terrestrial solar cells, and in the mid-1970s AEG-Telefunken started to make multicrystalline solar cells in collaboration with the Wacker company, which produced the wafers. Siemens also started an R&D effort with small-scale manufacturing in Munich. The RWE subsidiary Nukem initially tried to develop a copper sulphide solar cell but then switched to the MIS silicon technology. In France, in 1979, the Photowatt company started production as a Philips subsidiary in Caen.

It subsequently moved to Bourgoin-Jallieu and went through a number of changes of ownership, becoming part of the major French utility EDF in 2012. It remains one of the few terrestrial producers from the 1970s era still in operation today. Tideland Energy began solar cell production in Australia in 1981. BP became involved as a manufacturer in that same year through a joint venture with the LUCAS company. BP Solar went on to acquire Tideland and started solar cell manufacturing first in Spain and then in India. In 1999, after BP's merger with Amoco, Solarex became part of BP Solar, with the combined company becoming the largest global photovoltaics manufacturer in 2000.

The 1980s were boom years for photovoltaics start-ups. In addition to the ones already mentioned, many other oil companies were also active in the field. In 1973, Shell invested in copper sulphide cell production, but then switched to crystalline silicon, building up capacity in the Netherlands and Germany and acquiring the ARCO Solar facilities from Siemens. The Showa Shell company in Japan developed the CIGS technology. Mobil invested in EFG technology in the United States, while the Italian oil company ENI established the Eurosolare PV manufacturing plant. This was a period of great optimism, with major multinationals seeing a large business emerging and investing heavily in R&D for new materials, production, and marketing activities. In the early 1990s, however – before the German 1000 Roof Programme had its impact – some disillusionment set in. The anticipated low cost and high profitability had not been realized. Closures and rationalisation occurred. AEG was in financial difficulties, and after a series of joint ventures its assets merged with those of Nukem to form RWE Solar, which acquired the EFG technology from Mobil Solar as the parent withdrew from the market. Wacker terminated its solar programme, which was largely taken over by the Bayer Group, with Bayer Solar formed in 1994. For a period in the mid-1990s, there was very little solar cell production in Germany.

The market situation was changing rapidly, however. The success of the 1000 Roof Programme stimulated renewed interest and investment in the photovoltaics sector. Germany became the fastest-growing market, and the centre of manufacturing efforts. From its earliest years, photovoltaics technology had been exploited by major multinationals, each basing its manufacturing activities on in-house development, albeit with good support from research institutes. By the mid-1990s, a new type of company had emerged, which could take advantage of those offering turnkey solar cell manufacturing capabilities. The new companies were able to raise capital on stock markets and set up manufacturing facilities without large R&D activities. Solar World and Q Cells are good examples of this, both seeing stock market issues in 1999. SolarWorld acquired Shell's crystalline silicon assets in 2006.

The European and American manufacturers supported the rapid market expansion that occurred between 1997 and 2010, particularly in Europe (see Chapter 5). The annual world market was 100 MWp in 1997 and grew to 1.2 GWp in 2004, with 100% growth from the previous year. The 10 GWp annual market threshold was passed in 2010. While the 2000s were a boom period for European industry, manufacturing growth was hampered by the limited supply of semiconductor-grade silicon feedstock. This gave an opportunity for new manufacturers, especially in East Asia, to enter the European Market. In 2006, for example, there was a deficit of nearly 1 GWp from local manufacturers in Europe, which was met by imports from China, Taiwan, and Japan (see Figure 5.12). In 2007, Q Cells was the world's largest cell manufacturer, but it and First Solar in the United States were the only non-Asian producers in the top ten manufacturing companies by volume. The European market declined from 2010, and European and American manufacturers entered a period of severe price competition, primarily from China. Shell had largely exited the business by 2006. BP Solar ceased production in 2011, although its Indian manufacturing joint venture continued, wholly owned by the Tata Group. In 2012, Q Cells was acquired by the Korean Hanwha group, and it now operates mainly as a development facility. Schott Solar terminated its activities in that same year, ending the manufacturing begun by AEG in the 1970s. Solar World struggled on for some years but entered bankruptcy proceedings in 2017. This leaves Photowatt as the only major European solar cell manufacturer today.

As a consequence, the European industry has gone through a transformation, moving from solar cell manufacturing to the design and installation of large-scale grid-connected photovoltaic power plants. BP re-entered the photovoltaics business in 2017, acquiring a large share in Lightsource to form one of Europe's largest installation companies [5]. Shell has also started a programme of investment in renewable-energy technologies [6]. ENEL Green Energy, a subsidiary of the Italian utility, has installed 3.7 GWp of solar production facilities worldwide. The previous experience of these multinationals in the photovoltaic sector has provided them with the expertise to develop new business in a different sector of the market.

9.5 Role of China as a Photovoltaics Manufacturing Base

China was not a photovoltaic manufacturing nation prior to 2000. In 2001, a returning PhD student from the University of New South Wales set up a new company, Suntech, to manufacture solar cells there, producing 10 MWp in 2002 [7]. This was equal to the previous four years of photovoltaic production in the country. Suntech was very successful and obtained a New York Stock Exchange listing in 2005. This provided funding for expansion of the company, which became the world's largest photovoltaics producer by 2012, with 2 GWp of product.

The growth of large markets in Europe and the potential for a growing home market attracted investment and new start-ups. The Chinese Township Electrification Program for households began in 2003. By 2005, 13 Chinese photovoltaic companies had obtained international stock market listings and a further 11 were listed in China. These companies were able to take advantage of the gigawatt-scale market that already existed in Europe. As detailed in Chapter 5, studies had shown that photovoltaic module manufacturing costs below €1/Wp were achievable for manufacturing plants of

>500 MWp p.a. In addition to photovoltaic cells and module manufacturing, investment was made in upstream production of polysilicon feedstock and wafers. In 2012, a typical selling price for photovoltaic modules in Europe was €1.2/Wp [8], whereas modules from China were priced as low as €0.7/Wp [9]. This reinforced the trend from 2005 onward whereby European companies had not been able to produce enough to satisfy the local market demand. European markets also shrank after 2010 as FITs were cut back in response to the low prices. Widespread closures of European-based photovoltaics manufacturers occurred. The low prices from Chinese manufacturers led to claims of module dumping, and punitive tariffs were applied by the European Commission in 2013 [10] and by the US government in 2012 [11]. The European restrictions were lifted in March 2018, but they continue in the United States, with a 25% import duty. The price differential also continues, with current spot prices for Chinese modules in the German market at €0.32/Wp, as compared to €0.42/Wp for European ones [12].

The slowdown in the European markets after 2010 was more than offset by the growth of new ones, particularly in China, India, and the United States. In 2017, the Chinese internal market was 52.8 GWp – more than half of the total world market. This has enabled Chinese manufacturers to maintain their global leadership. Table 9.1 details the 10 largest photovoltaic manufacturers in 2018, as compared to 2008. It is clear that Chinese manufacturers dominate world production. The top five companies supply 60% of the global output [13]. Only one non-Asian company is now in the top ten: First Solar, supplying cadmium telluride thin-film modules.

Chinese companies have not neglected R&D. For example, Jinko Solar recently announced an independently verified 24.38% ($252 \, cm^2$) efficiency for its Cheetah PERC cell [14]. Given the strength of the Chinese market, the photovoltaic manufacturing infrastructure, and the economies of scale that Chinese companies enjoy, these are likely to be the lowest-cost commodity manufacturers and the main source of photovoltaic modules for some years to come.

9.6 Potential for Continued Market Growth

The global photovoltaics market has grown continuously since the early 1980s, with the output in 2017 at 100 GWp being 1000 times greater than that in 1997 at 100 MWp. The obvious question is whether or not this can be continued. There are clear drivers for growth, particularly in emerging economies, where additional energy is needed to fuel their economic growth. The challenge is to supply an estimated 434 million households not connected to a grid, as well as to improve the supply to those that are [15]. Within the energy market, electricity demand is growing strongly. It has been estimated that demand globally will double by 2050, despite efforts to improve energy efficiency, as the increase in electric vehicles increases electricity consumption [16]. This is illustrated in Figure 9.1.

Table 9.1 Ranking of the world's largest photovoltaics manufacturers for 2008 (Source Photon International) and 2018

2008			2018	
Rank	**Company**	**Country**	**Company**	**Country**
1	Q Cells	Germany	JinkoSolar	China
2	First Solar	United States	JA Solar	China
3	SunTech	China	Trina Solar	China
4	Sharp	Japan	LONGi Solar	China
5	JA Solar	China	Canadian Solar	China
6	Kyocera	Japan	Hanwha Q Cells	Korea
7	Yingli	China	Risen Energy	China
8	Motech	Taiwan	GCL-Si	China
9	SunPower	United States/Philippines	Talesun	China
13	Sanyo	Japan	First Solar	United States

Sources: Photon International and www.pv-tech.org/editors-blog/top-10-solar-moduel -suppliers -in-2018 accessed 10/06/2-19 [13]

Figure 9.1 Projected growth of global electricity demand 2000–2060, according to different bodies [16] (*Source:* World Energy Council: World Energy Insights Brief-Global Energy Scenarios Comparison Review 2019. © 2019, World Energy Council. Courtesy World Energy Council.)

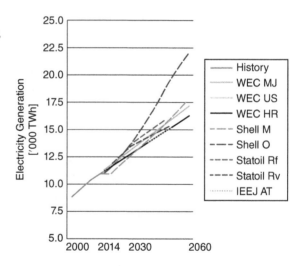

In the short term, it has been confidently predicted that world annual shipments of photovoltaic modules will rise to 160 GWp by 2022 [17]. The challenge in meeting the growth in Figure 9.1 is to do so without expanding the production of electricity from fossil fuels, in order that the UNFCC Paris Agreement to limit climate change to 1.5 °C by 2050 can be achieved. As has already been stated [2], this means effectively net zero greenhouse gas emissions by 2050. According to one study, this is attainable only if 85% of electricity production in 2050 is by renewables – mainly wind and photovoltaics [18]. A more detailed study has looked at the possible scenarios in the highly developed European economy. It concluded that if all available renewable resources were used realistically, including an ambitious 40% use of the available bio-mass, then 3.364 TWp of photovoltaics would have to be installed in Europe by 2050 [19]. Given the installed capacity in 2017 was 114 GWp [17], this would require a linear addition of 100 GWp p.a. to 2050. European installations were less than 10 GWp in 2017, and the peak year was 2011 with 20 GWp. If the global demand for electricity is to be met without catastrophic greenhouse gas emissions, all renewables must play a part, but given that photovoltaics have the highest resource and have the greatest geographical availability, they must take the greatest share. Present levels of installa-tion are well below those needed to meet the challenge [18]. Hence, the potential for growth is both urgent and very large.

9.7 Future Technology Development

While solar cell technology and manufacturing have made very strong progress since the first demonstration of a viable 5%-efficient silicon solar cell in 1954, all successful terrestrial solar cells have been single-junction. Good progress has been made in the development of thin-film solar cells, and laboratory efficiencies for both cadmium telluride and CIGS are over 22%, but these technologies have not displaced silicon. The only instance where the Schockley–Queisser limit [4] has been breached is in tandem III–V cells for space applications and concentrators, where cell cost is less significant. Silicon solar cells of over 22% efficiency are in commercial production at a number of companies. A practical limit for these is thought to be 25%, but this is low compared to the ultimate thermodynamic limit of 85.4% [20]. This clearly signposts that there is huge potential for further development.

The best single-junction GaAs cell has an efficiency of 29.1%, while the best four-junction tandem cell with wafers bonded has an efficiency of 46% at 500 suns [21]. These multijunction approaches are thought to be capable of achieving 50% efficiency in the medium term. They are high-cost technologies, and are unlikely to be commer-cially viable for large-scale terrestrial deployment. As described in Chapter 8, tandem cells on silicon offer the best prospect of achieving efficiencies exceeding 30% in a cost-effective manner.

Perovskite-based solar cells have emerged in the last 6 years and are still in an early stage of development. Scale-up to large areas has been challenging, and there are issues around stability and environmental concerns over the lead content. The perovs-kites do have the potential to exceed 30% cell efficiency as tandem cells, so long as materials development can provide a range of bandgaps from 0.8 to 2.0 eV while main-taining good photovoltaic properties. Perovskite tandems on silicon offer the best short-term potential, but much further development is required to demonstrate a robust, environmentally sustainable product. Good progress is being made in III–V on silicon without the disadvantages of the perovskites, but cost reduction remains a significant challenge. It is encouraging to note that given sufficient resourcing, technologies which were perceived as high-cost in the past can become cost-effective, such as the silicon PERC cell.

The subject of electricity storage is not within the scope of this book. Nevertheless, its importance cannot be neglected entirely. As already detailed, photovoltaics and other renewables will have to become the main technologies for electricity generation in the future in order to avoid the massive negative impacts of climate change. In some countries, particularly within the tropics, the solar resource is relatively constant through the year, but in northern altitudes the difference between winter and summer is very large. A typical study for the United Kingdom showed that the country's electricity demand could be filled by photovoltaics alone using only 1% of the land area – equal to the area covered by buildings [22] – but there was a massive seasonal shift, with the solar resource in a summer month being eight times that in a winter one [23]. There has been a huge increase in grid-connected battery storage for load levelling in developed countries, and not just for renewables. Grid-connected battery storage increased by 68% in the United States in 2017, to 1.3 GWhr [24]. However, this is used for levelling out day-to-day variations. Long-term storage for seasonal variation is a much bigger challenge. Where a country has a good hydroelectric infrastructure, pumping water into reservoirs using photovoltaic-powered pumps is a viable means of long-term storage [25]. Pumped storage has been used successfully for 25 years at the Dinorwig Nuclear Power Station in the United Kingdom. Not all countries have such suitable resources, however. Other solutions are needed.

While photovoltaics have become a large-scale industry, in technology terms it could be said that phase 1 has just been completed and a whole new generation of technology with efficiencies beyond 30% is still to emerge, together with the supporting storage to make it the globally preferred sustainable electricity-generating technology.

9.8 Final Analysis

This book have recorded the rise of an electricity-generating technology from interesting laboratory observations in the nineteenth century to a gigawatt-scale industry supplying an as yet small but rapidly growing portion of the world's electricity demand in an environmentally friendly and sustainable way. For much of this history, as detailed in Chapter 1, there was considerable scepticism that photovoltaics could be a commercially significant energy technology. It may seem inevitable that an energy source capable of meeting 6000 times the human population's energy demand and lasting for another 4 billion years should have been exploited by an energy-hungry planet, but success was not guaranteed. This book has described many of the challenges it faced.

To make the technology a success, the intervention of three entities was essential. First, there had to be the scientific discovery that light could produce electricity, and then the further development of a device capable of providing usable amounts of power. It took roughly 100 years for this phase to be completed, and it was very much dependent on the greater understanding and advance in semiconductor technology that took place during this time. Second, once the solar cell was shown to be useful, it took a generation of visionaries to realise its potential. The space satellite market provided the first application, which proved the effectiveness and durability of photovoltaic electricity generation. A brave band of entrepreneurs then set about creating a viable terrestrial photovoltaics market. This was a vital stage in bringing electricity to many previously off-grid communities around the world, increasing their health and prosperity to great social benefit. Simultaneously, electricity supply solutions were found for many remote professional applications. Finally, while this development brought considerable benefits, a transformation to a low-cost, high-volume globally significant energy business was needed. This required government action. While many governments supported photovoltaics R&D and demonstration programmes into the 1990s, it was only the introduction of the feed-in-tariff (FIT) in Germany that allowed it to become one of the lowest-cost electricity-generating technologies in the world, offering the potential for large-scale carbon dioxide-free electricity production into the future.

The initial driving force behind photovoltaics technology was the desire for energy independence and the avoidance of reliance on imported fossil fuels. With time, this focus diminished, but it was replaced by much greater concerns over greenhouse gas abatement and stabilisation of climate change. To go from the present scenario of 3% of global electricity production supplied by photovoltaics to the 85% required to minimise climate change will need another 10-fold increase in annual production. It is the role of governments to put the regulatory framework in place to continue these developments, while the research community must increase solar cell efficiency, reduce costs, and devise energy storage solutions. The manufacturing industry, meanwhile, must continue to expand capacity and make the investments necessary to meet the global need. The success to date has resulted from the cooperation of these three entities. The way ahead is utterly reliant on their continued mutual interaction.

References

1 IEA PVPS Annual Report 2019.
2 European Commission: 'Going Climate Neutral by 2050 – A Strategic Long Term Vision for a Prosperous, Modern, Competitive and Climate Neutral EU Economy' (2019).
3 F. Alt in 'Solar Power for the World' ed. W. Palz pub: Pan Stanford (2014) 301–306.
4 W. Schockley and H.J. Queisser: Journal of Applied Physics 32(3) (1961) 510–519.
5 https://www.bp.com/en/global/corporate/news-and-insights/bp-magazine/lightsource-bp-one-year-on.html accessed 7 August 2020.
6 https://www.greentechmedia.com/articles/read/shell-new-energies-director-on-investing-in-clean-energy accessed 7 August 2020.
7 H. Qin in 'Solar Power for the World' ed. W. Palz pub: Pan Stanford (2014) 427–436.

8 W.G.J.H. van Sark et al.: Proceedings of the 27th EUPVSEC (2012) 4384–4388.
9 G.P. Willeke and A. Raeuber: 'On the History of Terrestrial PV Development with a Focus on Germany' pub: Elsevier (2012) ch. 3.
10 Photon International 9 (2018) 34–35.
11 PV Magazine 8 August 2018.
12 Photon International 3 (2019) 45.
13 https://www.pv-tech.org/editors-blog/top-10-solar-module-suppliers-in-2018 accessed 7 August 2020.
14 Jinko Solar Press Release 3 June 2019.
15 International Finance Corporation: 'Off Grid Solar Market Tends Report 2018' (2018).
16 World Energy Council: 'World Energy Insights Brief – Global Energy Scenarios Comparison Review' (2019).
17 Solar Power Europe: 'Global Market Outlook for Solar Power 2018–2022' (2018).
18 IRENA: 'Global Energy Transformation – a Road Map to 2050' (2018) 10.
19 M.-C. Leonhard, M. Kamberaj, and H. te Heesen: Proceedings of the 35th EUPVSEC (2018) 2104–2108.
20 J.F. Guillemoles in 'Solar Cell Materials – Developing Technologies' eds G.J. Conibeer and A. Willoughby pub: Wiley (2014) 5–34.
21 M.A. Green et al.: Progress in Photovoltaics 27 (2019) 3–12.
22 T.M. Bruton and N.B. Mason: Proceedings of the PVSAT 8 (2012) 73–76.
23 T.M. Bruton and N.B. Mason: Proceedings of the PVSAT 7 (2011) 101–104.
24 W. Driscoll: PV Magazine 4 September 2018.
25 S. Bhattacharjee and P.K. Nayak: Renewable Energy 135 (2019) 288–302.

Index

Photovoltaics from Milliwatts to Gigawatts: Understanding Market and Technology Drivers toward Terawatts,
First Edition. Tim Bruton.
© 2021 John Wiley & Sons Ltd. Published 2021 by John Wiley & Sons Ltd.